Praise for THE BOOM

"Russell Gold's *The Boom* is a double quest. He tells the story of the biggest innovation in energy so far in this century—the shale gas revolution. He captures the personalities, and the drama and surprises, and brings clarity to the debate about the environmental impact—and what it means for the U.S. economy and 'energy independence.' But it's also a more personal story—about 'the Farm' in rural Pennsylvania where he spent time as a child, and his quest to understand what is happening in this new age of shale gas."

—Daniel Yergin, author of *The Quest: Energy, Security, and the Remaking of the Modern World* and *The Prize*

"In *The Boom*, Russell Gold provides a compelling account of the last half century of natural gas technology development. Driven by hunches, large ambitions, and even larger personalities, the story of fracking is the story of innovation, American style. Gold delves into the growing conflict between economic development and concerns over environmental damage, and explains why fracking is seen by some as a vital bridge to a sustainable energy future and feared by others as another excuse to postpone addressing the risks of climate change. . . . *The Boom* puts a human face on the unfinished story of our struggle to transition to a sustainable world."

—Steven Chu, former United States Secretary of Energy

"Gold's book is an early must-read for 2014: it is both a thorough and fascinating examination of the fracking economy and the technological innovations that have made these new riches accessible (including the often catastrophic damage done in the process of obtaining them)."

—*Gizmodo*

"An insider's guide to the most controversial energy-production technique in the United States."

—*Kirkus Reviews*

"Gold delivers an engaging and expansive education on the promise and risks involved with the sudden rise of fracking for oil and natural gas in the United States. . . . Gold delivers a balanced analysis weighing the benefits (the reduced use of dirtier coal, an end to the reliance on foreign oil and foreign entanglements, and sudden and reliable abundance of energy supply) against the pitfalls (the impacts on the environment and quality of life as energy companies stampede to secure leases and rush to drill, often in populated areas). Worthy of the attention of both fracking's boosters and opponents, Gold's insightful reportage supplies a well-rounded view of a polarizing subject."

—*Publishers Weekly* (starred)

"Whether you think fracking is our salvation or an agent of environmental destruction, *The Boom* is worth your time."

—Forbes.com

"Combining lucid explanations of fracking's technical aspects with the practice's more dramatic backstory, Gold's work is a tour de force of contemporary journalism that will captivate anyone concerned with the future of energy consumption and our rapidly changing climate."

—*Booklist* (starred)

"[R]eaders from both camps, and all the rest of us caught in between, will read *The Boom* and come away feeling like we've acquired a more nuanced understanding of an energy play that is changing

Texas, changing the United States, and changing the global energy economy."

—*Rivard Report*

"This deftly handled account of the shale revolution provides a sobering assessment of the current limits of alternative energy, making for a nuanced treatment of an issue too many would prefer to see in black and white. . . . Mr. Gold performs a valuable service by looking at it from a historical, economic, political and environmental perspective . . . his clear, thorough treatment of the subject is the starting point for a more informed discussion of energy and environmental policy."

—*Pittsburgh Post-Gazette*

"[*The Boom*] brings new clarity to a subject awash in hype from all sides. . . . A thoughtful, well-written and carefully researched book that provides the best overview yet of the pros and cons of fracking. Gold quietly leads both supporters and critics of drilling to consider other views."

—Associated Press

"*Wall Street Journal* energy reporter Russell Gold has produced a thoughtful piece of journalism, exploring the complex landscape of drilling, finance and politics that brought a gusher of oil and gas to a country convinced that its hydrocarbon heyday was over."

—*Nature*

"Russell Gold's *The Boom*, authoritative and fairly balanced, is a welcome guide—the best all-around book yet on fracking."

—*San Francisco Ch*

"[An] engaging story about the rise of fracking and how it has changed the energy landscape. Deep down, the book is a story about individual choices playing out against the wider energy landscape. . . . And, in the steady hands of Gold, a *Wall Street Journal* energy reporter and Pulitzer Prize finalist, the book ranges into a thorough explanation of fracking itself."

—*Austin American Statesman*

"*The Boom* marries the muscly prose of a beat reporter with a flair for finding compelling characters and telling anecdotes around this once-obscure oilfield technology."

—*Houston Chronicle*

"An in-depth look at the newest and controversial technique of extracting natural gas and oil."

—*Shelf Awareness*

"An excellent, fair-minded, engaging book. . . . Gold's words tell a dramatic and engrossing story. The book is well-informed and well-told: a great job of reporting."

—*Cleveland Plain Dealer*

"Gold brings clarity to a subject awash in hype from all sides. It's a thoughtful, well-written and carefully researched book that provides the best overview yet of the pros and cons of fracking."

—*Contra Costa Times*

"[A] revelatory and a cautionary tale . . . illustrates how dramatically America's energy equation has been rewritten in less than a decade."

—*Texas Monthly*

THE BOOM

How Fracking Ignited the American
Energy Revolution and Changed the World

RUSSELL GOLD

SIMON & SCHUSTER PAPERBACKS

NEW YORK   LONDON   TORONTO   SYDNEY   NEW DELHI

Simon & Schuster Paperbacks
An Imprint of Simon & Schuster, Inc.
1230 Avenue of the Americas
New York, NY 10020

This Simon & Schuster trade paperback edition May 2015

SIMON & SCHUSTER PAPERBACKS and colophon are registered trademarks of
Simon & Schuster, Inc.

For information about special discounts for bulk purchases,
please contact Simon & Schuster Special Sales at 1-866-506-1949
or business@simonandschuster.com.

The Simon & Schuster Speakers Bureau can bring authors to your live event. For more
information or to book an event, contact the Simon & Schuster Speakers Bureau at
1-866-248-3049 or visit our website at www.simonspeakers.com.

Interior design by Ruth Lee-Mui
Map by Paul J. Pugliese
Jacket art and design by FDT Design

Manufactured in the United States of America

10 9 8 7 6 5 4 3 2 1

The Library of Congress has cataloged the hardcover edition as follows:

Gold, Russell.
    The boom : how fracking ignited the American energy revolution and changed
the world / Russell Gold.
        p.   cm
    1. Petroleum industry and trade—Environmental aspects—United States.
2. Oil wells—Hydraulic fracturing.   3. Energy policy—United States.   4. Energy
consumption—United States.   I. Title.
    HD9565.G65   2014
    333.8'230973—dc23   2013028446

ISBN 978-1-5011-2261-3
ISBN 978-1-4516-9230-3 (ebook)

To my sources of energy:
Isaiah, Joaquin, and Laura

Contents

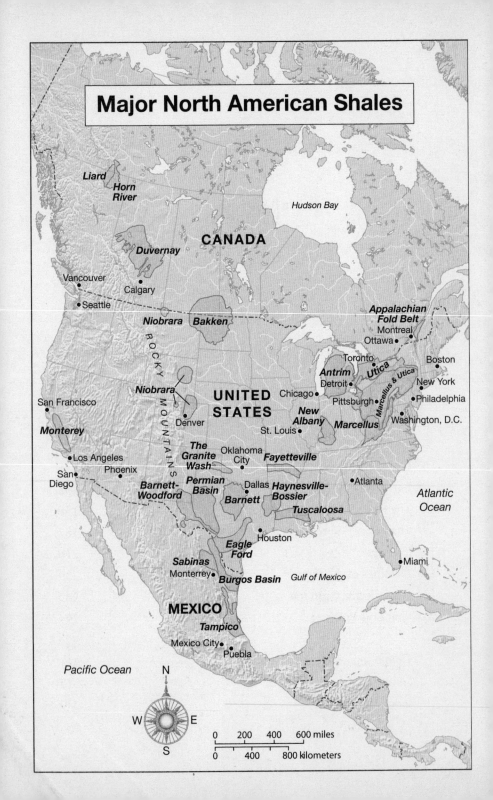

Major North American Shales

Liard

Horn River

Hudson Bay

CANADA

Duvernay

Vancouver

Calgary

Seattle

Niobrara   Bakken

Appalachian
Fold Belt

Montreal

Ottawa

Toronto

ROCKY MOUNTAINS

Antrim

Utica

Boston

Niobrara

UNITED
STATES

Chicago

Detroit

Marcellus & Utica

New York

San Francisco

Denver

St. Louis

Pittsburgh

Philadelphia

New
Albany

Marcellus

Washington, D.C.

Monterey

The
Granite
Wash

Oklahoma
City

Fayetteville

Los Angeles

Phoenix

Barnett-
Woodford

Permian
Basin

Dallas

Haynesville-
Bossier

Atlanta

San
Diego

Barnett

Atlantic
Ocean

Tuscaloosa

Houston

Eagle
Ford

Miami

Sabinas

Monterrey

Burgos Basin

Gulf of Mexico

MEXICO

Tampico

Mexico City

Puebla

Pacific Ocean

N

W   E

S

0   200   400   600 miles

0   400   800 kilometers

I recognize the right and duty of this generation to develop and use the natural resources of our land; but I do not recognize the right to waste them, or to rob, by wasteful use, the generations that come after us.

—Theodore Roosevelt, speech in Osawatomie, Kansas, 1910

There is only one rule of thumb in fracturing: that there are no rules of thumb in fracturing.

—M. B. Smith and J. W. Shlyapobersky, in *Reservoir Stimulation*, reference book for petroleum engineers, 2000

THE BOOM

1

JUST ADD WATER

A few years ago, my parents faced an unexpected choice.

Chesapeake Energy called them with an offer. The company wanted to drill for natural gas underneath 102 acres of land they owned with some friends in north central Pennsylvania. Would they be interested in signing a lease? The offer was $400,000 up front, plus royalties on any gas unearthed from the ground. It was an astounding amount of money for terrain so rocky and hilly that the local dairy farmers didn't want it.

Despite a name that evokes sailboats and seafood, Chesapeake hails from landlocked Oklahoma City. Once little more than a two-person partnership, it grew to drill more wells than any other company in the world. At its peak, it held leases to punch holes in an area the size of Kentucky. Its annual budget topped $20 billion in 2012, and it spent a chunk of that on a sophisticated advertising campaign that preaches the gospel of domestic energy production and attempts to calm fears about hydraulic fracturing. Chesapeake drills more than a thousand wells every year and fracks each one. Once the bit churns through the dense rock, the company pumps in millions of gallons of water and chemicals to create a network of sinewy fractures, each one an escape route for trapped hydrocarbons. Gas and oil freed from the shale flows out of the cracks and up the well. The recipe that Chesapeake said it was following was quite simple: just add water. The reality, however, was more complex.

My parents' property was valuable to Chesapeake because it sits atop one of the largest shale formations in the world. The Marcellus Shale was once so obscure that it appeared in only the most detailed geologic maps of the area. The charcoal gray rock runs from New York, crosses Pennsylvania, and stretches into Ohio and West Virginia. Near the Farm, as we call the property, it is more than a mile deep. Over millennia, the shale cooked at just the right temperature and pressure to turn long-dead microorganisms into trillions of cubic feet of natural gas. Chesapeake wanted to extract the gas and sell it to households and power plants. Sometime in the future, when my parents turn on their stove or television in Philadelphia, about 180 miles to the southeast, some of that energy might have begun its journey a mile beneath their property.

The size of Chesapeake's offer shocked my parents, but not me. I was then—and still am—an energy reporter for the *Wall Street Journal*. In the mid-2000s, my beat was the "independents," a group of midsized companies that didn't sell gasoline or operate refineries. They drilled for oil and gas in the United States. A scrappy bunch, they were the descendants of the industry's wildcatter heritage. They didn't have the money and engineering muscle to compete with globe-straddling energy titans such as Chevron and BP for giant projects in the Middle East or deepwater exploration off the African coast. The independents fed on the table scraps of the energy feast.

But the picked-over United States held a surprise. By the beginning of the twenty-first century, little-known independents had figured out a way to get natural gas from dense slabs of buried shale rock. I reported on this new phenomenon and wrote the first national newspaper articles about the gas around Fort Worth in the geologic formation known as the Barnett Shale. I watched the transformation of the suburbs, as Little League fields turned into tawny rectangular drilling pads. I had a front-row seat as this energy upheaval ranged from Texas to Arkansas and Oklahoma and then vaulted to North Dakota and Pennsylvania. Using fracking, the independents found an unbelievable amount of gas—and then oil as well. Early optimistic estimates of how much was available turned out to be absurdly conservative. Even Exxon Mobil, the embodiment of the modern energy behemoth, began to look for a way to get involved.

Not long before the energy industry beat a path to my parents' doorstep, I traveled to Pittsburgh to write an article about the leasing and drilling in southwestern Pennsylvania. This early drilling took place hundreds of miles away from the Farm. Fracking was spreading farther and faster than I had realized.

My parents and their friends bought the Farm in 1973 and built a small house. It was a place to get away from Philadelphia for a weekend, a couple hours up the Pennsylvania Turnpike but a world away from their brick row house and busy city life. Back in the 1970s, they were immersed in the left-wing, antiwar politics of the day. For the first few years, they called the property Oriente, after the Cuban province where Fidel Castro and Ernesto "Che" Guevara began their revolution. From their first days as landowners, these urbanites stood out in this conservative and poor part of the state.

While their neighbors in Sullivan County worked long hours to wring a living from their acres, the white-collared Philadelphians kept their land untouched. Making money from the land wasn't in the plan.

The trust documents reflect their vision. There were eight owners. Each owner, or couple if married, would pay an equal share. If a couple wanted to sell their share, they wouldn't profit. They would get back the money they put in to buy the property and anything paid over the years for taxes and upkeep. Drafting the trust in an era of antigovernment protests, they figured, when the revolution came in the United States, they could always escape from the chaos to rural Pennsylvania. In the more likely scenario, in their minds, of a government crackdown on radical dissidents, the house could be a way station on the way to exile in Canada. And if none of this Armageddon came to pass, it would be a place for inexpensive vacations, where their city kids would have a chance to run around the woods and swim in a pond.

That anyone would want to drill wells on the land in search of natural gas was beyond the realm of imagination. This wasn't Texas. When my mother called me to discuss the offer, she wanted to know what I thought. Should they sign the lease? It is a complex question, and answering it requires weighing sacrifice and opportunity, money and the environment. As a reporter, I spend my working hours talking to people who work in the industry and live near its wells. I think about how much energy the world consumes and where it comes from. There are no easy answers to the energy puzzle. There are unforeseen costs and necessary evils.

What the independents set in motion has changed an entire global industry and upended the traditional energy order. The emergence of vast, untapped energy stores, literally under our feet, allows natural gas to challenge coal and nuclear power as the dominant fuel used to make electricity. It opens the door for renewable energy to emerge as a force in its own right. But it also extends the age of fossil fuels for decades, a profound challenge to the climate. The revolution had come, after all, but it wasn't the one my parents feared in the 1970s. It also wasn't in Philadelphia. It was on the Farm in Pennsylvania, and in metropolitan Fort Worth, northern Louisiana, frigid North Dakota, and rural Ohio. The revolutionaries also weren't disaffected Philadelphians, they were geologists and petroleum engineers from Texas A&M University.

This revolution is transforming the United States. To a remarkable extent, this once-obscure oil-field technique defines the nation's economic and environmental future. Fracking has unleashed more oil and natural gas than anyone thought possible. It is providing an abundance of domestic energy, helping to drive a rebirth of manufacturing, and easing dependence on overseas energy peddlers. Accessing this energy requires tens of thousands of new wells, each fracked with enough water to fill several Olympic swimming pools and hundreds of gallons of chemicals. It also requires turning whole counties into industrial zones, complete with fleets of trucks, air quality concerns, a disruption of nature, and fear that water aquifers will be poisoned.

Modern societies run on fossil fuel. There is a direct connection between the number of jobs, cars, factories, and computers a country has—in short, its economic prosperity—and its energy consumption. Every day, the world consumes ninety million barrels of oil. Nearly one of every five of those barrels slakes the thirst of the United States's economy and commuters. America is the most affluent nation in the history of the world, and it consumes more per person than any other major country ever has. Oil—and its main product, gasoline—has become a birthright of modern industrialized economies. We pull into gas stations and expect there to be enough gasoline to fill our tanks. Gasoline is everywhere, but it is invisible. It flows out of the pumps, through thick synthetic rubber hoses, and into our cars. You can smell it and occasionally see a hazy vapor. But you rarely, if ever, see it. Where does it come from? Not from the gas station on the side of the road. That is its last, brief stopover in a long journey.

Much of this energy comes from overseas. Without thinking about it, we have exported the dirty work of finding and developing oil fields, along with the environmental and social costs, to other nations. Until a few years ago, we planned to do the same for natural gas. But this dynamic is changing. Increasingly, crude oil consumed in the United States begins life in places such as North Dakota and South Texas. Fracking allows America to produce the gas we need— and much of the oil also—in our backyards. The promise and peril of energy production is coming home. The traditional energy system is being torn down and rebuilt. It's an opportunity to take a hard look at the energy we use.

In 2008 a small Canadian energy advisory firm issued a report titled *The "Shale Gas Revolution."* The name stuck and is now used widely, mostly by supporters of this new energy production who want to emphasize how big and pervasive the changes are. I also refer to it as a revolution, but for different reasons. It is a revolution because the old order is tumbling. King Coal's reign as the nation's predominant fuel for making electricity is tenuous, and even petroleum's stranglehold on powering vehicles is weakening. As with many revolutions throughout history, once change is set in motion, the end result can be unexpected. Revolutions also create their own stories, creation myths, and hagiographies, as well as boogeymen.

This book tells the story of fracking and how it rose from a minor oil-field tool to a world-changing technology. It is also an attempt, amid the tumult, to dispel some fictions that have risen to accepted "fact."

The Farm isn't part of my world anymore. When I hit my teenage years, spending a weekend with my parents and older sister had become excruciatingly boring. Completing thousand-piece jigsaw puzzles in a house without a television didn't cut it.

As I was stumbling toward adulthood, in the 1980s, two men in faraway Texas and Oklahoma were going through their own changes. In time, they would help propel shale rocks from obscurity into the topic of boardroom presentations in the highest echelons of American capitalism. George Mitchell was a most unusual Texas oilman: liberal and an early convert to sustainable development. He created the Woodlands, north of Houston, to showcase that building a new community didn't require bulldozing all the trees. At the same time, the eponymous Mitchell Energy & Development was one of the largest oil and gas companies in Houston, the world's energy capital. Its most important holding was a gas field around Fort Worth.

Mitchell geologists noticed that every time their wells passed through shale rock in search of conventional pockets of oil and gas, instruments registered a significant gas presence. There was fossil fuel in the rocks, but it was as inaccessible as the sword in the stone from Arthurian legend. Mitchell's long wells could reach the gas, but the company's engineers had neither the tools nor the knowledge to get it out. Open up a textbook from that era and look up how to drill a well into shale and, if it mentioned the rocks at all, its advice was to look elsewhere. But in 1982 Mitchell Energy drilled the C. W. Slay #1 well to target the gas trapped inside the Barnett Shale, a thick geological formation that covers five thousand square miles, fanning out from Dallas to the west and south. Though the company had fracked wells in the past, it had never tried fracking shale rock. It worked, sort of. Gas flowed from the shale. But it was expensive. As a wildcat well, it was underwhelming. But as a science experiment, it showed promise.

Through the 1980s and into the 1990s, the company drilled a couple wells into the Barnett Shale each year. Mitchell's engineers kept chipping away at this rock, trying to figure out how to force the shale to give up its gas. They pumped in heavy, gelatinous liquids they hoped would muscle their way in. Then, as they were ready to give up, a young engineer came up with a simple and elegant solution to cracking open the rock that would make these shale wells both less expensive and more bountiful. It was a new approach to fracking that used more horsepower and employed water, the Earth's most abundant liquid. It was the beginning of the revolution. By then, Mitchell was nearly eighty years old. At the time, his children weren't interested in the oil field, and he wanted to sell his company. But the rest of the industry remained skeptical about his shale wells. Wasn't this new technique just a ploy by aging management to hype the company and get a buyer to pay top dollar?

When Mitchell was first trying to crack the shale puzzle, a different oilman was starting out. In 1981 Aubrey McClendon returned to his hometown of Oklahoma City after attending Duke University. Oklahoma City was in the midst of an energy boom. Global events led to a doubling and then a tripling of oil prices. He came home to prosperity, Cadillacs, and new skyscrapers. But he wasn't a geologist or an engineer. He was an aspiring accountant who had graduated from college magna cum laude with a degree in history. He entered the energy industry and soon became a landman. His job was to convince landowners to sign leases to allow rigs to drill for oil and natural gas on their property. In 1982 a global recession led to a swift collapse in crude prices, and the city's banks reeled from aggressive oil loans. The local Penn Square Bank failed. It was the first of more than one hundred Oklahoma bank failures. Bankruptcy auctioneers replaced those Caddys as the city's unofficial symbol.

It must have been quite an education, unlike any that Mc-Clendon had received at Duke. He witnessed the boom-and-bust nature of oil and gas. He saw the riches available if you could time the rise and fall of volatile commodities correctly, and he also saw how money made all this possible. In time, he would go on to found Chesapeake Energy and become a convert to the potential of shale gas. He would do more than anyone else to promote shale gas. He was part pied piper, part early adopter, and part rapacious capitalist. Those dense rocks that resemble an old-fashioned chalkboard would make him a billionaire, before he nearly lost it all. McClendon would use his energy wealth to advance his energy and political agenda, assemble a world-class wine collection, and uproot the Seattle SuperSonics of the National Basketball Association, bringing his hometown its first professional sports franchise, renamed the Oklahoma City Thunder. More than anyone else, he would usher in an era of energy abundance.

History is full of odd ironies. The birth of shale gas is no exception. An environmentally minded oilman, George Mitchell, pioneered a way of cracking open rocks with water and chemicals that would come to embody one of the greatest environmental fears of the twenty-first century. And a right-wing oilman, Aubrey McClendon, would become an outspoken prophet for an abundant, low-carbon source of energy.

Not long after Chesapeake inquired about leasing the Farm, my father spent a day driving around to visit neighbors and discovered that many had signed leases already. The reality sunk in. Future drilling locations surrounded the Farm. "We believed they would go under our property and get the gas anyway," he told me later. It is an old fear. At the beginning of his classic novel *Oil!*, Upton Sinclair captured how the industry played on this worry. "Take it from me as an oilman," the budding tycoon J. Arnold Ross tells a group of neighbors. "There ain't a-goin' to be many gushers here at Prospect Hill; the pressure under the ground will soon let up, and it'll be them that get their wells down first that'll get the oil." This race to drill and drain free-flowing reservoirs was how it worked at the beginning of the twentieth century, but it is no longer the case. Still, the fear remains.

My mother called me again. "It is going to happen, and it is going to be obtrusive," I said. But it wasn't necessarily all bad, I added. Gas was a low-carbon energy source. By signing the lease, she was contributing to its growth. Until you sign the lease, you have the upper hand, I told her. They want your land. Craft an agreement that gives you a say over where the wells will be drilled to keep them on the periphery of the property, out of sight. It was possible for the industry to coexist with the land. In 2004 I visited Ted Turner's ranch in New Mexico's Sangre de Cristo Mountains. It is a spread so beautiful, it was once considered as a possible national park. Turner allowed gas drilling but wrote a lease that contained stipulations to make sure there was minimal impact. The energy company could bring only so many trucks onto his property at any one time. The wells were camouflaged behind low walls. The company tried four different shades of paint before Turner's ranch manager settled on one that blended with the ponderosa pines. I suggested that my parents take a similar approach.

The Farm's owners met in March 2009. "There was the inevitability of change," my father recalled. "It was coming. All of our neighbors, all of the land around us, had signed. We were really concerned we would get all the negative—the trucks, the noise—and none of the positive stuff: the money."

My parents and their friends signed the twenty-page lease in October 2009. In January 2011 the Oklahoma City company drilled the Matt 2H well at the cattle-guard gate to the property, fracking it in August. The Farm's owners had become partners with Aubrey McClendon. Did these left-wingers, now balding and with gray hair, members of the Philadelphia upper middle class, do the right thing? These days, my sons have started going to the Farm every summer with their grandparents. Was the choice to sign the lease going to change the land? Or, by signing the lease and throwing their lot in with fracking, were my parents helping secure a future for their grandchildren filled with low-carbon gas and renewable energy?

There is a story my parents like to tell about the summer of 1973, as they were building the house that has become part of the lore of the land. A truck arrived and unloaded the house on pallets. None of the Philadelphians had any experience with building or hard labor. Three were newly minted lawyers and community organizers. My mother was getting ready to attend medical school. They gamely threw themselves into the job. My father volunteered to handle the wiring. He had taken electric shop in seventh grade, but when a county inspector arrived, he pronounced the wiring out of code. My father had used electrical boxes that were too small. Small bribes were the way to navigate Philadelphia's bureaucracy, so he figured the Sullivan County inspector wanted a payoff.

As the inspector walked away, my father trotted after him with $50 in his pocket. Rambling, he described how things were done in Philadelphia. The inspector understood the message. "You see those little kids back there?" he said, pointing to a two-year-old boy and a three-year-old girl. "If anything happened to those kids because of the wiring, I would never forgive myself." The inspector was talking about me and my sister. He walked away, the $50 still in my father's pocket.

The inspector had the right approach. If you are going to build something, make sure it is safe. He refused to sign the papers until the wiring was done right. My father replaced the electric boxes. A few days later, the inspector approved the job. The United States faces a similar challenge. We are tearing down the old energy order and building a new one, but are we doing it responsibly? Is it enough to be passive consumers of energy, turning on lights and turning up the thermostat, and relying on the energy industry to make sure the electrons and gas molecules are there for us? What does it mean to promote ethical energy production, and how can it be done?

After my folks signed the lease on their land, I struggled with a set of nagging questions as I traveled the country, digital recorder and steno pads in my backpack, talking to people in fracking hot spots. Had my parents made the right choice? Had I given them good advice? Was the nation making a horrible mistake, or were these energy executives ushering in a new era of energy that we can all embrace? Or are they extending the lease on life of fossil fuel, energy that is both wonderful and destructive?

One thing is certain. Nearly every well drilled in the United States is fracked. That's one hundred wells a day, perhaps even a bit more, year-round. Whether you fear fracking or celebrate it, that's a lot of holes in the ground.

This book is about the ecosystem and inhabitants of the new United States, one that I sometimes call Frackistan. To trace its emergence, I will begin deep underground and follow the path of the hydrocarbon up and out of the rocks. Humans do not create crude oil and natural gas. We gather it from deep underground, where it is created. Any book about fossil fuels must begin with rocks. They are, literally and figuratively, the foundation of the entire story.

Before the rigs are assembled, a company acquires the right to drill a well. This often involves finding who owns the mineral rights. As was the case with my parents, it's the landowners. But in some cases, the mineral rights have been severed from surface ownership. One person owns the land and someone else owns what's underneath. This split can be problematic for all involved. The hunt for leases is a central element of the story of modern American energy.

Once leases are signed and wells drilled, the energy molecules enter a labyrinthine system of pipes and machinery built by the energy industry to extract the oil and gas. To begin this journey requires fracking, the violent act of cracking open rocks. Without this initial interaction between humans and rocks, there would be no resurgence of US energy production, no fracking, and no book. I will spend time with the people, beginning more than a century ago, who pioneered fracking. Moving upward, the story becomes about the wells themselves and the freshwater in aquifers near the surface.

Of course, the story doesn't end at the surface. That is where the energy industry interacts with people: neighbors who live near the wells, government officials, and environmentalists. It is also here that chief executives and corporations set this activity in motion and interact with the Wall Street money machine, without which the wells never would have been drilled. But this is not the end of the journey, and this book will also trace the final step. Eventually the bulk of this energy is burned to create electricity or heat homes. This releases carbon dioxide, which heads upward into the atmosphere and contributes to climate change. All along this path, the book will spend time with many of the inhabitants of this land, many of whom have struggled with the complexities of this new era and have arrived at surprising conclusions.

Why was Chesapeake so keen on north central Pennsylvania, in and around land my parents bought four decades ago? Why had a gaggle of landmen encamped at the Sullivan County Courthouse, rifling through real estate records untouched for decades to find who owned the mineral rights and bewildering overworked clerks who felt like the circus had come to town? The energy industry was chasing a giant deposit of sedimentary rock called the Marcellus Shale.

How did all this energy get here in the first place? Let's begin at the beginning, or, at least, a long time ago. Most of the world's continents have mountain ranges in the middle and great coastal plains on the edges. North America is different. Sixty million years ago, the broad collision of plates created the Rocky Mountains in the west and the Appalachian Mountains in the east. In between was an enormous, shallow ocean called the mid-Cretaceous inland sea. It covered the area we now call the Great Plains, Texas, and even reached up into what became Pennsylvania. Zooplankton and other small aquatic organisms lived in that sea, fed by the sun and nutrient-rich waters. When they died, they settled on the seabed. In this vast marine environment, over millennia, these dead creatures created a thick layer of organic material. Eventually rocks buried this sediment, an overburden that created pressure and generated heat. The organisms slowly cooked, broke down, and turned into natural gas and oil. Petroleum geologists have a simple term for shale. It is "source rock"—the birthplace of oil and gas. If water is the most abundant ingredient in fracking, it is also the ecosystem that generated the best energy-laden shale rocks in the first place.

Shales formed all over the planet as landmasses shifted around, leaving these fossil-fuel generators scattered on every continent and underneath the oceans. But not all shales are the same. Ancient lakes created shales that tend to have waxy hydrocarbons that cling to the rocks and resist modern petroleum extraction. Elsewhere, such as in China, where forest and woody debris constituted the primary organic ingredients, the resulting shales are layered with silty rocks, like a kitchen sink full of haphazardly stacked dirty dishes. But in the great waterway that covered North America, the conditions were just right for generating large, contiguous shales—the kind of deposits that in the modern era attracted Chesapeake, because it could carve up entire counties into rectangular units, each with its own set of wells, and drill with confidence that ninety-nine of every one hundred would find natural gas trapped inside the rock. It was an ideal setup for a shale factory.

At the end of the Cretaceous period, the waters receded, and the North American continent dried out. Beneath the surface, it was chockful of source rocks. The first that caught the attention of Mitchell Energy petroleum engineers in the early 1980s was the Barnett Shale in Texas. Driving from the northernmost well to the most southern would take about three hours depending on how heavy traffic was passing through downtown Fort Worth. By mid-2012, rigs had drilled more than fifteen thousand Barnett wells, mostly in four or five counties close to Fort Worth. They will drill thousands more before energy companies can no longer get enough gas out of the ground to justify the cost. And even then, gas will continue to seep out of the rocks and into wells for years after the drilling rigs have moved on.

In a global context, the 5,000-square-mile Barnett Shale is on the small side. On the other side of the world, Russia's Bazhenov Shale sprawls for about 850,000 square miles, from the frozen Kara Sea nearly all the way to the steppes of Kazakhstan. It is about the size of Texas and the Gulf of Mexico combined. Due to a lack of roads, it is impossible to drive from its northern to southern boundaries. By one estimate, long-gone rivers deposited eighteen trillion tons of organic material on the bottom of an ancient sea. That's a lot of carbon and hydrogen molecules that ended up as organic-rich siliceous shale.

This is the source rock for Siberia's giant oil fields—and the source of much of modern Russia's wealth and political strength. The trillions of tons of dead organisms baked in the heat of the earth and transformed, slowly, into oil droplets. Oil was expelled from the shale and traveled upward until impassable rock canopies trapped it. This process formed large oil reservoirs, not in underground pools but inside permeable rocks riddled with tiny holes that allowed billions of barrels' worth of the sought-after liquid to collect. In any given year, Russia is either the world's largest, or second largest, oil producer. This is due primarily to oil that escaped the Bazhenov. This narrative is a quintessential story of fossil fuel: organic material is converted into oil and gas inside the shale, and then the hydrocarbons escape and travel upward. If something blocks the molecules' path, they will fill up porous rocks like water in a sponge and create a petroleum reservoir.

Almost every one of the world's giant hydrocarbon reservoirs has filled up with molecules that were formed in shale source rocks and then exited at such a glacial rate that a time-lapse camera set to snap a frame every decade would make for boring viewing. California's thick molasses crude came from the Monterey Shale. In North Africa, the Silurian shales of Algeria and Libya have earned those countries membership in the oil-exporting cartel OPEC

(Organization of Petroleum Exporting Countries). Farther east, the same shales generated gas trapped in carbonate rock cavities now known as the giant North Field between Iran and Qatar, the world's largest gas reservoir. In northern Europe, the Kimmeridgian Shales led to the 1970s North Sea oil boom. The poetically named La Luna Shale sits under Venezuela. The Qusaiba "hot shale" is believed to be the source of Saudi Arabia's Ghawar, the largest single collection of crude oil that has ever been—and likely will ever be—discovered. These are conventional reservoirs that until a few years ago were the exclusive target of the world's petroleum industry. "Drill a well and drain the reservoir" was the oilman's mandate. But a century of rising global thirst for oil and gas has begun to exhaust these warehouses. The market demanded more, and the industry responded. It knew the nursery rocks still contained oil and gas, but how much? And could it be coaxed out?

The answers that have emerged over the past decade have spurred the industry forward. Many shales leaked off most of their hydrocarbon wealth, but others kept theirs locked away. Some, such as the Bazhenov, are so large that they leaked off billions of barrels but still have billions more stored away. How many more Barnetts and Marcelluses—geological turnkeys—are there? We don't know. Until recently, few bothered to ask. The conventional thinking was that shales did not hold much economic value, and funding to study these rocks was paltry. Juergen Schieber, a professor of geology at Indiana University and one of the few academics interested in shale before the energy industry discovered how to frack it, said the gaps of knowledge remind him of sixteenth-century maps with large empty spaces and "Here be dragons" notations.

Humans have been following a path toward the source rock for as long as history has been recorded. At first the only hydrocarbons used by our ancestors made the long trip from source rocks up to the surface, avoiding geologic dead ends. Humans reveled in the utility of this gift. They used it to caulk boats and cook food. In places, it thickened the earth and was shoveled out to bind bricks together. In ancient Mesopotamia, the builders of the Tower of Babel and the Hanging Gardens of Babylon likely used this oleaginous bitumen as mortar. Eventually humans dug deeper in search of more. In 1821 a shallow well in Fredonia, New York, produced enough natural gas to light streetlamps for the town. In Europe and Central Asia, oily dirt was mined and refined into kerosene to be burned for light. As demand for the fuel grew, tinkerers and then professional engineers invented ways to delve into the earth in search of larger, untapped reservoirs. Refiners distilled this crude oil into fuel. Before long, there was the Ford Mustang and men on the moon.

Civilization has been heading toward source rock for a few thousand years; toward the geological kitchen where heat and pressure turned organic material into hydrocarbons. Now, with fracking, there's a lot more oil and gas to be extracted. But once we've reached the source rock, we've gone as far back as possible. You can't devise technology to dig deeper and reach even further back in geologic time. Source rock is where plankton turned into hydrocarbons. There is no further back. This is it.

2

OTTIS GRIMES

Drilling and fracking is a loud, noisy business. Smelly too. People rely on energy and enjoy its benefits, but who wants to live near a well? Sometimes the petroleum-laden rocks are conveniently remote. In 1967, drillers discovered Alaska's enormous Prudhoe Bay oil field a couple hundred miles north of the Arctic Circle. Years later, floating drilling platforms tapped into massive oil and gas deposits in the deep waters of the Gulf of Mexico, a long helicopter flight from shorefront homes. The shale boom is different. Wells spread out across entire counties. As of 2013, more than fifteen million Americans lived within a mile of a well that had been fracked in the past few years.

This new proximity between wells and homes is one of the defining features of the new energy landscape. If wells weren't allowed within a mile of the nearest home, United States oil and gas output would very likely be declining. But that isn't the path we've chosen. The country and its courts long ago decided that unfettering drillers provided desirable rewards. Some homeowners would be compensated for their troublesome new neighbors; others weren't so fortunate. This arrangement is being tested in the era of fracking.

To better understand the complex relationship between drillers and homeowners, why it was codified, and how it enabled the birth of fracking in the United States, it helps to go back to where it started.

In March 1919 an itinerant oil driller named Ottis Grimes bought a house for his family in Burkburnett, Texas. Until a year earlier, there were a thousand residents in this town near the Red River in the northern part of the state, along with a bank, hotel, and cotton gin. Then a local farmer drilled a well and found a lot of oil.

Burkburnett's metamorphosis was sudden and complete. Wildcatters, land speculators, and oil-field workers turned the small prairie community into a bustling hub of humanity. By the time Grimes arrived, the sidewalks were too packed to walk down. Restaurants welcomed patrons around the clock. Several dance halls opened, and then were closed for "immoral" activities. There was even a drug bust. Sheriff's deputies confiscated $42,000 worth of morphine sulfate—the equivalent of about three-quarters of a million dollars when adjusted for inflation.

The population quintupled in a couple years. It was a quintessential boomtown. The 1940 Hollywood movie *Boom Town,* featuring Clark Gable and Spencer Tracy as wildcatters-turned-tycoons fighting over the same woman, was inspired by Burkburnett.

The transformed city bore all the hallmarks of early oil production. "The air was permeated with a strong smell of petroleum," a historian noted. Hundreds of ten-story derricks, built from longleaf yellow pine, towered over the city. "Burkburnett is now like a huge pin-cushion," a visitor remarked. Derricks were often only fifty feet from each other.

Traffic and oil spills turned the city's dirt roads into a fetid mess. A black-and-white photograph showed a black creek running between homes. "Oil a mile long," was etched into the print. The month before Grimes bought his home, two British insurance syndicates canceled liability coverage in Burkburnett. Their reason was the town was so low on water that fighting the frequent well fires was impossible. An influenza outbreak was followed by smallpox.

The toll on the town was matched by the opportunity for wealth. In November 1919, the local paper reported an oil company out of Dallas paid a record $100,000 to drill on a single acre. The local school leased its land for $14,000 and built a new high school with the money. The entire city was gripped by oil fever. "To try to talk about anything except oil," said a journalist, "would create as much consternation as a rebel yell at a spiritual séance."

The Burkburnett boom was different from other early-twentieth-century discoveries in one important respect: the oil was found in and around—and under—a town that had been platted and subdivided into hundreds of small homesites. The oil field was not on a remote ranch. The oil field was Burkburnett. Many people received leasing payments and royalty checks. It was the first time that energy production had landed, quite literally, in people's backyards. Or front yards, as was the case with Ottis Grimes.

He bought his home, a four-room frame house on a small lot, but there was a catch: the transaction didn't include the mineral rights, creating what's known as a "split estate." Grimes owned the land and the house. H. L. Bunstine, who worked for Magnolia Petroleum (later Exxon Mobil) and sold him the house, kept the oil underneath.

Within days after the sale, armed Goodman Drilling workers

entered the property. They erected a soaring derrick that Grimes said posed a "constant menace" of toppling in the strong winds and flattening his home. The shaft of a several-hundred-pound steam engine spun throughout the night within inches of his stoop, and a steam boiler was deposited behind his home. Workers built a "slush pit" to hold noxious drilling fluid so close to the house that windows were spattered. Grimes and his wife couldn't sleep. They had to yell to talk inside their living room.

Grimes sued to stop the drilling, claiming that Goodman Drilling had turned his domesticity into a nightmare. The company's lawyers replied that any delay would allow nearby wells to suck out all the oil. A considerable number of the company's employees would need to be laid off. Anyway, this was Burkburnett, the lawyers argued, and "the injuries complained of . . . are but those commonly sustained by the inhabitants of the town." As many as four wells were being drilled on any given day in Burkburnett and if the court granted the requested injunction, they said, the oil-greased wheels of commerce would stop turning.

A nearby judge ruled in favor of the drillers, and Grimes appealed to a court in Fort Worth, where other judges also said he was out of luck. He owned the surface, but not the mineral rights beneath his property. "He is in no position to complain of conditions," the judges wrote, "such as are usual and customary during the drilling of an oil well."

After losing his court fight, Grimes and his wife sold their house and left the oil patch. They moved to the Texas-Mexico border and raised dairy cows, according to census data and land records. The legacy of the case, however, outlived his short stay in Burkburnett and influenced the modern energy boom. The Grimes vs. Goodman Drilling ruling set the legal framework in Texas, and for decades turned up as precedent in other oil and gas exploration cases.

"The general rule has been pretty much the same since the Grimes case," said Barney Fudge, who grew up in Burkburnett in the 1940s and serves as a judge in the 78th District Court, where the Grimes case was originally heard in 1919. "Whoever has the mineral rights is the dominant estate. Texas did that because the hydrocarbons were so valuable. I think it was a policy decision by the courts."

Other states copied Texas's legal approach. The oil industry had the right to drill wherever it owned the minerals. And not just drill: build roads, tear down fences, put in pipelines, and use water wells. The dominance of mineral rights was an invaluable boost to the fledgling oil industry. Eager for the benefits of energy production, politicians and judges created favorable conditions for the United States to become a giant oil and gas producer in the twentieth century.

Over the years, Texas and other state courts eased up. By the 1970s, the courts tilted back a bit toward landowners. Oil companies could only use what land was "reasonably necessary" and consistent with typical industry practices. Today's courts and state regulators are considerably less tolerant of wanton disregard for landowners and the environment than they were in 1919. But the principle remained in place as shale development shifted into high gear in the first years of the twenty-first century. In legal terms, the landowner remains the servant while the mineral owner, and the companies that lease these rights, is the master.

This decidedly pro-drilling legal framework is one of the reasons fracking was an American invention. The right conditions existed in the United States to encourage oil and gas exploration and the risk taking necessary to propel the industry forward.

There are other reasons as well. From Saudi Arabia to Mexico, in Europe and Africa, oil and gas belong to the state. State-run energy companies, in many parts of the world, administer and exploit these national resources. The governments usually own all the oil and gas, even if they are under private property. The United States chose a different path. It has never had a national oil company. American colonists rejected English common law, which reserved all mineral rights for the monarch. Initially, landowners in the United States owned their minerals and this created an enormous incentive for them to allow oil and gas drilling, because any wealth went into their wallets. If someone other than the landowner held title to the subterranean riches and stood to reap all the profits without any disruptions, the incentive would be even greater. America's private ownership of mineral rights, conceived while whale blubber, coal, and wood were still the fuels of choice, turned out to be remarkably useful in the petroleum age. It was "a marvelously elegant system that ensures that all natural resources are fully developed," enthused Rex Tillerson, the chairman and chief executive of Exxon Mobil.

A permissive legal system, large financial incentives for the owners of mineral rights to allow drilling, and a tradition of small, independent energy companies struggling for survival and willing to take risks created an environment where fracking took root and flourished. From the dribble of gas in Mitchell Energy's 1982 C. W. Slay well, it took twenty-six years for the fracking industry to reach an annual production of 1.84 trillion cubic feet of natural gas from shale. It took two years to double that to 3.68 trillion cubic feet. And it took less than two years to double that again to 7.36 trillion cubic feet.

Like it or not, fracking is here to stay. What began in Texas moved to neighboring states and then across the country. It is now spreading around the world. The sun never sets on a frack crew. Fed by a steady diet of fresh capital from investors, the drilling industry proceeded with abandon, not caution. This headlong rush created a glut of gas and reversed decades of declining oil production. This flush of fuel created new wealth, jobs, and economic opportunity. The phenomenal growth of fracking took everyone by surprise. The energy industry wasn't prepared, and neither were landowners and government officials. While fracking upended the energy landscape in many ways that are beneficial, it also had its own set of problems. "This came much faster than anticipated," said Peter Voser, the chief executive of Royal Dutch Shell, in an interview. "And neither the regulator, the legislator, nor the industry was actually prepared to deal with the issues."

Today's boomtowns bear only slight resemblance to Burkburnett nearly a century ago. There is still a rush of wealth and jobs, housing shortages, and often a surge in drug use. Landowners with mineral rights are generally ready and willing to sign large leasing bonus checks. Modern-day Ottis Grimeses, who live amid the trucks and diesel-powered compressors but don't have a share of the prosperity, tend to be the loudest critics. John Tintera, who for twenty-two years worked at the Texas oil and gas regulator and retired in 2012 as executive director, said his biggest mistake was not to recognize the problem of surface owners who had all the nuisances without getting any compensation. In a speech in September 2013, he said he never heard from the people "getting money in their mailbox on a regular basis . . . the real complaints were from surface owners." There will likely be more latter-day Ottis Grimeses who live with drilling but don't get checks. Developers and sellers are increasingly holding on to the mineral rights, hoping that drilling may one day begin.

But there are many significant differences between Burkburnett and modern shale communities. Gone are the mile-long creeks of flowing oil. Reckless environmental degradation, at least in the United States, often results in fines and criminal convictions. Energy production near where people live has brought about more community involvement, accountability, and lawsuits. The industry is being scrutinized more closely than ever. States are playing catch-up, struggling to become fleet-footed regulators with the backbone to stand up to industry. Many remain conflicted, however, and want to make sure they don't choke off the economic gains created by drilling.

Emily Krafjack is trying to find what she calls the "delicate balance" between the benefits and headaches of drilling. She lives with her husband in a rural home in Mehoopany, Pennsylvania. She leased her property, as did her neighbors. Chesapeake Energy chose to build a drilling pad on the edge of her neighbor's property, which put it right on the edge of her small plot of land. It drilled the first well five hundred feet from her porch in 2010.

"It was a lot louder than we expected. Everyone told us we would hear a hum. I would have loved that. We heard every clang and bang and every worker yelling," she said. She slept, or tried to, with the television on in a futile attempt to drown out the noise. Her husband, a construction worker, told her it was louder than a pile driver. Once, a convoy of trucks filled with sand backed up in front of her home, filling it with diesel fumes. "I thought I could grin and bear it, but I was coughing my head off," she said.

Unlike Grimes, she didn't sue. She appreciated the jobs the companies brought into her community. "I have friends and family who got jobs. Many of us have benefited from bonus payments and royalties. It can be very good for our local economy," she said. (Her observation is borne out by national statistics. Between 2010 and 2012, the United States added 169,000 fossil fuel–related jobs, a pace ten times quicker than the rest of the economy.)

Krafjack educated herself about drilling and created a nonprofit organization to spread information about how to hold the industry to the highest possible standards. She learned that some companies built sound barriers around their drill pads to muffle the noise, and insisted Chesapeake build one. It did for the second well. "I don't think I'm being unreasonable. I can't talk on the phone or watch my television when they are fracking," she said. When she called Chesapeake and complained about the trucks, they were gone within an hour. "There are times when the delicate balance is reached. When an operator or a pipeline company decides to do something better, they are working towards the balance," she said.

Mark Boling, a top executive with Southwestern Energy, a Houston driller, agreed there is more work to do. "The industry has done a great job of figuring out how to crack the code belowground—how do you get natural gas or oil out," he said. "However, it hasn't spent a lot of effort thinking about how you handle development aboveground."

Living near a well under construction isn't easy. Traffic, noise, and foul air are constant and legitimate complaints. A Chesapeake vice president, in a candid speech, said in March 2012 that while people like natural gas, "making it can be problematic. Nobody likes that part. I can tell you this: the sausage making will get better and better and better."

The Burkburnett boom—and dozens like it—paved the way for the American century and its unmatched prosperity. Will the shale boom, occurring in a new century with modern environmental sensibilities and concerns, resemble the old booms? Or is there a new path available to the United States, one that emerges from the new proximity of hundreds of thousands of shale wells and millions of Americans? Will the industry create a new generation of Ottis Grimeses—or will it create more Emily Krafjacks?

One thing is for certain. Simply to keep oil and gas production in the United States flat, the industry must drill thousands of wells every year, often packed closely together. This is due to the stingy nature of the rock.

Shale is dark and dense, surprisingly heavy in the hand. Scott Tinker reached into a three-foot-by-three-foot cardboard box and pulled out a cylinder of shale cleaved in two pieces. "Here's good, dark shale," he said. "The organics make it dark when they rot." A blackboard hue indicated that organisms once lived and died there, turning into fossil fuel. We were standing in a warehouse on the north side of Austin, Texas, piled from the floor to the thirty-foot ceiling with racks of these boxes. It reminded me of the final scene from *Raiders of the Lost Ark*. Tinker called his collection the Library of Congress of rocks.

Tinker, the head of the University of Texas Bureau of Economic Geology, one of the world's foremost institutes for studying petroleum rocks, held the shale near his face and licked the flat side of half a cylinder. His tongue cleaned the dusty surface. "Now you are looking at the rock texture," he said, holding it out for me to examine. It appeared uniform, with no visible striations, and solid. To see the tiny holes where the oil and gas is trapped requires a $2 million scanning electron microscope. Breaking into these vaults demands a lot of muscle.

The first step is to drill a long well straight down, which is then typically turned so that the hole, known as a wellbore, runs parallel to the surface, traveling through the horizontal layer of shale. A conventional sandstone reservoir, with large interconnected pores full of fossil fuel, is like an inflated balloon. When the balloon is punctured, air rushes out. Similarly, when the drill bit churns its way into a reservoir, the oil and gas stampede into the well. But shale is so solid that next to nothing enters the well unless it is "stimulated."

A company pumps in liquid—mostly water, mixed with sand and a cocktail of chemicals to reduce friction, thicken the water, and kill any hitchhiking bacteria carried from the surface—under extraordinary pressure. Water doesn't compress, so when forced up against a rock at rising pressure, it will cause the rock to break.

When it breaks, the frack fluid rushes out of the well and into the newly opened space. The fluid often carries sand, which will remain behind to prop open the new fractures and prevent them from closing up again when the liquid is retrieved. More water, more pressure, more fractures. A driller might execute dozens of fracks along the horizontal leg of a well, transforming an impermeable block into rock riddled with tiny cracks. Each crack exposes the shale to what amounts to a tiny brook that leads to a small stream and so forth until it reaches a river (the well itself) that connects the shale to the surface. A petroleum engineer who calculated how large an amount of shale was exposed by a typical frack estimated it was one hundred million square feet—or the floor-space equivalent of about thirty-five giant malls.

A lot of natural gas, or oil, will rush into the new fractures. A newly fracked well produces much of its energy in its first few months. But the story doesn't end here. There's an enormous pressure difference between the shale and the fractures. Think of opening the door of a plane at thirty thousand feet. Everything inside the cabin that is not bolted down will be sucked toward the opening, as the unequal pressures inside and outside find a balance. Something similar happens in the shale. Tiny gas and oil molecules can travel, possibly several feet, from inside the shale to the new cracks. Thus over months and years, a fracked well will continue to produce declining amounts of hydrocarbons as the molecules jump from one tiny pore space to the next, and likely through the rock itself, in a slow journey to the well.

Still, a single well drains oil and gas only from a few hundred feet around the hole. That's why the industry places wells close to one another, or else they will leave valuable hydrocarbons behind. This is not Burkburnett's haphazard, wasteful, a-derrick-every-fifty-feet approach, but it can still leave a county pockmarked. Researchers are searching for new ways to create bigger fracture networks. Today's fracking is like the tiny crack created by a pebble kicked up by a tire hitting the windshield of a car. Engineers are looking for ways to hit the windshield with a hammer, smashing it into a mad spider's web of fractures.

In the early years of the fracking juggernaut, the industry barreled ahead, eager to capitalize on this newfound resource. In its wake, it generated concerns and some real problems for people who lived near wells and for the environment. It would take years for these problems to be identified. Citizens, regulators, and some parts of the industry are hard at work trying to find fixes.

Fracking injects a large amount of water, and much of that water is put down a well, never to return. Less well understood is that the sheer volume of water that comes back out of wells means that contaminated water, not oil and gas, is the industry's largest product. What flows out needs extensive treatment before it can be reused in another frack job, much less released back into nature. Initially, some of this water was sent to public wastewater treatment facilities, which weren't equipped to handle it, and then discharged into rivers. Pennsylvania, where this problem was particularly acute, instituted a voluntary moratorium for municipal plants. Problem solved, until several private wastewater treatment facilities sprang up to take the waste, setting off a game of regulatory whack-a-mole. Some of these facilities are excellent operators, others less so. A study in Western

Pennsylvania found one private facility was releasing a large amount of salt and radium. Levels of the radioactive element did not exceed government standards, but radium was building up in some sediment near the plant's discharge pipes. Some of the returned water, instead of being cleaned up in a sewage plant, was simply injected into deep disposal wells, where it fills up underground reservoirs in the hope that it will never be seen again. Recycling frack water has been on the rise for several years, but is far from universally embraced.

These disposal wells, even overlooking the inherent problem of taking drinkable water and sending it into permanent exile, can lubricate existing faults and cause earthquakes. "Man-made seismicity" and "induced seismicity" are the polite terms for this phenomenon. While the tremors aren't large, they are nonetheless unsettling for longtime residents of Dallas and Columbus, Ohio, who are growing accustomed to feeling small rumbles under their feet.

Getting the fresh water to a well requires considerable truck traffic. For the Matt 2H on my parents' property, about 350 trucks delivered water from the Susquehanna River, which is thirty miles away. Another 75 trucks traveled sixty miles from Tunkhannock Creek, and 10 trucks from nearby towns. More trucks hauled in a small factory of equipment needed to drill and fracture the well.

So far, the shallow aquifer that supplies drinking water on my parents' property and neighboring farms hasn't turned briny. Tests for contaminants have been negative. The well itself descended 7,200 feet. If it was built correctly, it is doubtful that man-made fractures created any new pathways for existing pollutants, or chemicals used to create the well, to rise up into the aquifer. It isn't impossible, but independent studies in neighboring counties called such an occurrence "unlikely" and haven't turned up any evidence of this taking place.

If the fracking didn't create any pollution pathways, perhaps the well itself did. The layers of cement and steel pipe inserted into the wellbore and designed to protect shallow drinking water aquifers don't always work. My father asked for information about what tests Chesapeake ran on the well to determine if the cement held, but he gave up asking after getting the runaround for a couple weeks. Even if the well was good on day one, what will happen to the Matt 2H after a decade, or decades, of the pipes and cement sitting in a hot environment with corrosive liquids? Was Chesapeake in a rush to finish the well and move on to the next? I don't know. State inspectors cited the company for failing to follow "best management practices" to protect a nearby stream. What the inspectors saw on the surface led to violations, but what about what was below and out of sight?

There's another question about the well—and many others like it. When Chesapeake completed the Matt 2H, flushing out the water it injected underground, how much natural gas was allowed to vent into the atmosphere? How much escaped from the pipelines that delivered it around the country? Methane, the main ingredient of natural gas, is a potent greenhouse gas. Releasing it adds to the carbonization of the atmosphere and contributes to climate change. Indeed, if too much natural gas leaks out, the benefits of making electricity from gas versus coal can disappear. I have no way of knowing how much methane escaped from the Matt 2H. When future wells are drilled on the pad on the Farm, federal rules that go into effect in 2014 will require "green completions" to capture this methane. Some companies already use this equipment and report that not only is most of the methane captured, this approach often pays for itself by avoiding the expense of buying the diesel otherwise needed to fuel drill-site machinery.

In July 2011, a month before fracking of the Matt 2H began, the lead author of a large study on natural gas told the US Senate

Committee on Energy and Natural Resources that there are benefits and risks associated with fracking. The risks, said Ernest Moniz, then a professor at the Massachusetts Institute of Technology, are "challenging but manageable. In all instances, the risks can be mitigated to acceptable levels through appropriate regulation and oversight." As he said in a speech a year later, "All of these are manageable, which should not be confused with managing them." Three months after he made these comments, Moniz was nominated to become secretary of energy. Both the federal government and state governments are taking steps to identify problems and devise fixes. The environmental impact is improving. But more can be done and, as long as regulators and the industry don't shy away from problems, can continue to be done.

Thanks to fracking, the United States is producing more natural gas than ever. The same technology used to get gas out of shale is now being used to get oil as well. In the summer of 2013, the United States pumped nearly 7.5 million barrels a day of crude oil, a level unseen since 1990. North Dakota, home to the Bakken Shale, was producing 875,000 barrels a day, up from 150,000 barrels five years earlier. For decades, the United States has imported millions of barrels of oil every day. Imports are now falling. While it seems unlikely that America will ever become "energy independent," it is certainly unwinding its dependence on foreign suppliers in the Middle East and Africa. For generations, the United States has used its military might to keep oil flowing, fighting wars and patrolling sea lanes. Maybe this era will now come to an end. By 2020, America could become the largest global oil producer.

This is a stark change from the past, when so much energy production was outsourced. US energy consumers were able to use vast amounts of oil and gas without having to confront the impact or legacy of their addiction. A few years ago, I flew over Nigeria's coastal mangrove swamps in a helicopter. The vegetation on the sides of the Niger River was blanched white and denuded of leaves. It felt like I was looking at a black-and-white film. After decades of oil development, and the extraction of more than $1 trillion worth of oil, Nigeria struggles with rampant corruption and internal conflict. It remains a poor nation by any measure. The Niger Delta was the source of nearly 8 percent of the oil imported by the United States from 1981 to 2011. In the first half of 2013 it was 4 percent and falling.

Fracking means that the United States is producing more and more of the energy it consumes in its backyard. That didn't work out too well for Ottis Grimes. But there is no question that the localized environmental impact is significantly less than in 1919 Burkburnett, or modern-day Nigeria.

Drilling wells in a more environmentally responsible manner isn't enough for some fossil-fuel critics. They argue that burning coal, oil, and gas is releasing too much carbon and accelerating the unpredictable consequences of climate change. But not all fuel is equal in this measure. To achieve the same amount of energy, burning coal generates 42 percent more carbon dioxide than crude oil, which itself generates 18 percent more than natural gas. Critics argue that slowing the rate at which carbon is building up in the atmosphere, by burning gas instead of coal, is a half measure, and the Earth is too far into climate change for this kind of incremental progress. What is needed, they contend, is a wholesale switch to fuels that don't emit *any* carbon.

Sometimes I wonder what the energy landscape would look like if the industry couldn't frack shale rocks; if all that oil and gas were still locked away out of reach. Would there be more wind and solar power? Would we be putting liquefied corncobs and prairie grass into our fuel tanks? Would we have found ways to be more fuel efficient, investing in public transportation and insulated window-panes?

I suspect that America would be importing huge amounts of natural gas from overseas along with the fleet of tankers that brings crude oil. Fracking has not derailed the growth of renewable energy. Electricity from renewables, power sources that emit no carbon, has grown fast. The wind provided three-tenths of 1 percent of US power a decade ago. It is now about 4 percent. Considering that American energy consumption is Brobdingnagian, that's historic growth. Solar is also growing but remains much smaller.

Energy systems change when something better and cheaper comes along. New England's whalers stopped harpooning when refined petroleum proved a better light source. Crude oil replaced coal in trains and boats in the first couple decades of the twentieth century because it was a more compact, capable fuel. Natural gas is making inroads because fracking allows it to be abundant and cheap, at least in the United States. Some believed that the twenty-first century belonged to renewables—and it might yet—but fracking has breathed new life into fossil fuels.

Fracking is a challenge to renewable energy systems such as wind farms and solar arrays. The glut of inexpensive gas has made it hard for renewable energy to compete on the nation's power grids. But this competition is also forcing wind and solar to get better and, arguably, helping accelerate the maturation of these technologies. The abundance of inexpensive natural gas helped cushion the sticker shock of higher-priced renewables. Wind and sun are power sources that turn on and off. The wind doesn't always blow, and clouds can block the sun. For the power grid to function as we expect, there needs to be something to back up this intermittent renewable energy. Coal and nuclear power plants aren't well suited for the job. They don't like to be turned on and off quickly. But natural gas power plants are more nimble and can hold together a mélange of renewable and fossil fuels. This possible future is on view in Texas, which leads the nation in wind power and natural gas production. One study, paid for by renewable energy advocates, concluded the path to low-carbon power generation will require both gas and renewables working together.

The global energy system is vast. It won't change quickly. It is the foundation of modern life, and due to climate change, it poses a threat to modern life. The energy unearthed by fracking is both a once-in-a-lifetime opportunity and an enormous challenge. The rise of fracking is a story of ambition and resourcefulness. It is a tale that could occur only in the freewheeling United States, a nation of an enormous energy appetite, with no discernible policy on providing that energy, and a willingness to turn its back on Ottis Grimes. The Earth is warming, and once the source rock is depleted, the era of fossil fuel will end whether we are ready or not. Fracking has changed the energy industry and is changing the world around us. It is here to stay.

3

EVERYONE COMES FOR THE MONEY

A few miles south of Killdeer, North Dakota, amid rolling grass-
land that stretches out of sight in all four directions, two dozen
men huddle close together to hear Josh Byington over the blus-
tery prairie wind. "Welcome out here, gentlemen," says Byington,
his hooded eyes peering around from under a hard hat. Everyone
but him wears identical dark blue coveralls with silver reflective
stripes around their arms. Each has an American flag patch on the
right arm and the North Dakota state flag on the left arm. Some
workers left their hard hats unadorned except for the name of their
employer. Others personalized their hats with stickers. One guy had
"Frackn8r" on his hat. Another: "Coon Ass."

Byington is youthful and trim. Only when he takes off his hard
hat (sticker free) do you see that his sandy hair is being gradually
overrun by white. Just shy of forty years old, he has two decades'

experience in the oil fields of North Dakota, Wyoming, and Colorado. At this prefrack safety meeting, his word is final. Talk to the workers whose shift is ending, he urges. Learn if there been any problems. When we're pressuring the pipe to test it, don't stand near the iron. Be careful. Be smart.

"Pay attention to what is going on. This is big business. Hit the wrong switch, it could be your life or someone else's." He pauses. "No one gets hurt today, right?"

There are murmurs of assent. One worker yells, "Exactly!"

"All right, gentlemen," Byington says as he looks around, "let's go have some fun."

I have come to North Dakota to observe the fracking of the Irene Kovaloff 11-18H, a well on the southern edge of the Bakken Shale. There is nothing exotic about the well. It is one of a crowd of one hundred wells that will be fracked in the United States on this particular day in October 2012, ten in North Dakota alone. I could have chosen any number of places to witness a frack. But North Dakota offered an unvarnished view of the industry at full throttle. Trucks crowded the roads. In nearby Dickinson, where I stayed for a few days, the city engineer had resigned a week earlier, citing his increased workload and stress. That's what happens when five thousand new residents move into a city in two years, making one of every four Dickinsonians a recent arrival. In the midst of a sluggish national economy, locals advertised five openings for every person seeking a job.

Energy companies have come to North Dakota because when they frack the Bakken, light sweet crude oil comes out of the rock. Other shales offer mostly natural gas, which by the middle of 2012 was so plentiful there was a glut. So the rigs and the frack crews migrated to the Bakken. The industry is moving so quickly that the pipe layers can't keep up. Trucks haul the crude to rail depots, where it is loaded onto railroad cars. The wells here produce a bit of natural gas also, and no one wants to wait around for a connection to a gas pipeline, so they flare it off. At night, the onyx sky flickers with gas being burned off. One night I drove out from Williston, another oil boomtown, in search of a flare and found one within five minutes. Flames roared in a pit on the side of the road like a giant, unattended bonfire.

Over my journalism career, I have visited dozens of wells. But I had never spent time at the well as it was fracked. I felt a bit like a baseball fan who scanned the box scores and followed the standings closely but had never watched a game. I wanted to take my seat and settle back to watch the whole nine innings. Getting a ticket, however, wasn't easy. I called several companies active in the Bakken. They turned me down. One told me it was too busy to have its people take time off to "babysit" a visitor. Another said it just wasn't interested. Finally, Marathon Oil, a Houston corporation that generates an annual profit larger than the gross domestic product of some small countries, agreed to let me spend time on a frack job. The decision went all the way up to the chief executive. He decided that the industry had been hurt by its secrecy. When I got the okay, I went out and purchased a pair of steel-toed work boots.

A few weeks later, I waited in a North Dakota hotel room. The prairie winds had been near hurricane strength for an entire day, causing operations to come to a standstill. When my cell phone rang, a Marathon official told me that the winds had died down enough to allow the fracking of the Irene Kovaloff. I hopped in my rental car and headed north from Dickinson to meet my Marathon handlers.

Josh Byington oversees the movements of the couple dozen coverall-clad workers. His directive is simple enough: force more than one million gallons of liquid into the Irene Kovaloff under enough pressure to crumple the toughest car Detroit can turn out. Until his replacement arrives later that night, he makes sure the cocktail of water, sand, and chemicals was mixed just right. He checks that every flange is in place and every bolt secure. On this cold October day, the Bakken rock underfoot is the same thick, impermeable layer that it has been for the past fifty million years. By the time Byington has slept for a few hours and returned for his next shift, a full twenty-four hours after the fracking began, this section of the Bakken will be filled with thousands of tiny fractures, smashed into pieces like a shattered dinner plate. A week later, the valves atop the well will be opened, and crude oil will begin to seep through these new networks into the well. It will be pumped to the surface and, after being processed in a refinery, will end up as gasoline in automobile tanks and fuel in the airplanes overhead. By then, Marathon will have fracked several more wells.

The Irene Kovaloff, a recently drilled, long, and narrow hole in the ground, went straight down two miles until it reached the Bakken Shale, and then made a gradual 90-degree turn and continued for another two more miles in a southerly direction. Once fracked, the Irene Kovaloff dribbled forth a thousand barrels a day of crude for the first few weeks. Then the oil flow started to decline. By itself, the Irene Kovaloff is a drop in the global bucket. But Irene isn't alone. Nearby is her sister, the Viola Koberstine 34-7H. And not far away are relatives and neighbors: the Willard Kovaloff 21-17H, the Darcy Dirkach 14-12H, the Louie Hendricks 24-20H, the Wm. and Agnes Scott 14-25H, and the his-and-hers wells, the Tom Steffan

21-27H and the Deanna Steffan 44-22H. And so forth and so on across the prairie. Compare the list of well names to North Dakota homesteaders, and there is significant overlap.

Before workers transform the Irene Kovaloff into a producing oil well, they must make sure everything is secure and ready. When the prefrack safety meeting ends, a class system among the workers becomes visible. Most of the men head in one direction, where they check water levels in the rows of blue trailers, stand amid clouds of silica dust to make sure the sand doesn't fall off the conveyor belt, or tend to the dozens of machines. They spend their shift smelling musty diesel and tie their hoodies around their faces to keep out the cold. It is in the high thirties, but the constant wind makes it feel significantly colder. A smaller group heads in another direction and climbs a few steps into a temperature-controlled trailer called the data van. They spend their shift inside, peering out of thick-paned windows at the well, giving orders through headsets. They take off their hard hats and strip down to T-shirts under their coveralls. There is a coffee urn at their disposal. A car stereo bolted into a wall panel keeps them entertained during overnight shifts. On a recent night, Pink Floyd's *The Dark Side of the Moon* accompanied a frack. The next night it was Ted Nugent. The guys outside hope to work their way into a data van job.

Byington heads inside. He is at the top of the pyramid, and the only person not wearing coveralls. He is the "company man," the oil company's top representative on a job site. But Marathon Oil contracts out this work, and even though Byington is Marathon's man on the frack site, he works for StimTech, a company in Rock Springs, Wyoming, that provides a variety of skilled workers and well services to the industry. His job title is consultant. Marathon has invested $9 million to lease the location as well as drill and frack the Irene Kovaloff. The company usually has one employee at the frack site, responsible for health and safety. This arrangement is the modern corporate approach to oil production. The company subcontracts out nearly everything and leaves almost nothing to discretion. For this well, engineers in Houston created a "prog," a forty-page document that provides step-by-step directions. Workers will frack the well thirty times. Each frack has six distinct steps. It's all in the prog: when to switch chemical recipes, when to release four different gradients of sand, how much pressure to use. All of these instructions were uploaded into data van computers. "Marathon cooks this up, kicks it up to us, and we try to execute it to a tee," says Byington.

Inside the van, several workers with large headsets sit in gray swivel chairs bolted to the floor. They work at a long desk, with computers and monitors hanging from the ceiling displaying pressures, volumes, pH balances, and a dozen other measurements. There are four large windows in front of them looking out onto the industrial tableau of a large blue blender truck feeding the frack fluid into six giant trailers with silver pumps, all leading to a single four-inch pipe that rises to the top of a twelve-foot stack of red valves. The computers, the headphones, and the focused faces make the van feel a bit like a NASA command center. But rather than clean-cut engineers, the workers are scruffy. There is an assortment of beards and mustaches, and hair spilling out from under baseball caps. The North Dakota oil field has boomed so quickly that certain basic necessities—an appointment with a barber included—are hard to find. If a NASCAR pit crew had been hired to work at NASA, it would look a lot like the folks in the van.

A week ago, red stones covered this five-acre pad. The drilling rig had come and gone, leaving behind a long hole into the ground.

Now the Irene Kovaloff must be fracked, or else the oil will remain in the Bakken rock formation.

People have been drilling into the Bakken for decades. A Tulsa company called Stanolind Oil and Gas drilled the first successful well. This was in 1953—and a couple hours' drive north of the Irene Kovaloff. Stanolind fracked the Woodrow Starr #1 well one time with 4,900 pounds of sand and 120 barrels of crude oil. It produced 536 barrels of high-quality crude a day for four years until problems with the well caused the company to plug it up. Modern Bakken fracks are of a different scale. It's like comparing a pocket calculator with an eight-digit LCD display to a modern desktop computer. Byington orchestrates a job that used four trucks full of sand and more than one million gallons of water, which sit in sixty blue shipping containers that formed a wall around two sides of the pad.

What has changed in the nearly sixty years between the Woodrow Star and the Irene Kovaloff wells? The obvious answer is that the frack jobs have grown larger, more sophisticated, and more expensive. But there's another, less obvious, difference. For decades, companies have drilled a handful of wells a year into the Bakken. Some were quite good producers, especially those in parts of the basin where natural fractures and folds allowed oil to accumulate. Adding up all the output, the Bakken produced about one hundred thousand barrels of oil a day. The day the fracking of the Irene Kovaloff finished, the North Dakota Department of Natural Resources reported that oil production had topped seven hundred thousand barrels a day for the first time. Most industry forecasters expect it to exceed a daily output of one million barrels and perhaps reach two million barrels before leveling off. North Dakota, quipped oil economist Phil Verleger, "should start considering applying for membership in OPEC." At the time, the Roughrider State produced more than OPEC member Ecuador. It passed another member, Qatar, in the summer of 2013. It is fun to imagine North Dakota's governor, Jack Dalrymple, who grew up on his family's wheat farm near Fargo, nibbling on chocolate Sacher tortes at OPEC's Vienna headquarters, discussing oil quotas with ministers from Saudi Arabia, Venezuela, and Angola.

"There are three things that make a good well. Location, location, location," explained Pat Tschacher, Marathon's superintendent of well completions in the Bakken, when we met in Marathon's new Dickinson offices a couple days before the frack job. "The challenge we face is to make a good well in a bad location." The Irene Kovaloff is in a bad location. The Bakken here resembles a gray sidewalk. Picking up a piece of it, it is inconceivable that there is oil inside, much less that the oil can be extracted. But the industry figured out in the late 2000s how to make good wells in bad locations. That's what turned the Bakken into a giant oil field, where companies such as Marathon are fracking a few thousand wells a year. At each, the playbook is the same. Assemble a drilling rig and drill a well. Disassemble. Then bring in the frack equipment: "Sand King" trucks carrying four hundred thousand pounds of fine sand, conveyor belts, pipes, pumps, and chemical vats. Frack the well. Move on. Repeat.

At the Irene Kovaloff site, the weather disrupted this assembly-line efficiency. In the days before Josh Byington's speech, the moveable factory was brought in on dozens of tractor trailers. Machinery was carefully put in place until this patch of open ground resembled a crowded parking lot. Water trailers were backed into place until there was barely enough room to fit an arm in between them. Flatbeds with opaque tubs of chemicals were positioned near a dispensary truck. The giant sand trailers backed up on a conveyor belt that led to a hopper. Walking around the site requires vigilance to avoid tripping over the pipes and data wires jumbled underfoot. As the crew finishes setting up, the winds become gusty. Promotional bumper stickers boast that the industry is "Rockin' the Bakken." Now the data van is rocking, buffeted back and forth by the winds. A flagpole attached to the van's roof blows off its mooring. Moving the final pieces of heavy pipe with a truck-mounted crane is out of the question. On the nearby two-lane highway that connects Killdeer and Dickinson, tumbleweeds the size of German shepherds go marauding past.

After this delay, the wind abates enough to install the final pipes. But as Byington peers out the windows from inside the van a few minutes after the meeting, he sees another problem. "We've got a leak on number four," he says. During a routine prefrack pressure test, a compact mist of water was spraying from one of the six pumper trucks. Sly Henderson, a wiry man with short-cropped hair and a thin mustache riding atop his upper lip, looks out from inside the van and spotted the leak. "Rafael, can you please bleed the line?" he says into his mouthpiece. He issues orders in short, staccato bursts in a quiet, almost polite voice. The six pumper trucks are parked in a row, with only a couple feet between each truck and its neighbor. A blue-clad worker places one foot on the metal wheel guard of two trucks, straddling the open space. He delivers a dozen hard blows with a large hammer to the faulty valve to tighten it.

"Thank you, gentlemen, let's go ahead and clear the line," says Henderson. But when the equipment is pressured up to 6,800 pounds per square inch, the leak reappears. Henderson sighs. "Let's replace the gasket," he says. He picks up a Styrofoam cup and spits tobacco juice into it.

Waiting for his crew to loosen the valve and install the new gasket, Henderson explains that he ended up in North Dakota after his landscaping business in suburban Baton Rouge, Louisiana, tanked. "The economy went down like *this*," he says, waving his hand in a steep dropping motion. Unemployment was rampant, and for the first time in years, keeping his family fed and under a roof became a challenge. He was accustomed to working hard outside. His search for work brought him to Baker Hughes, a large oil-field service company. For two weeks at a stretch, he lives in one of the many man camps that have been built in North Dakota. Cobbled together in a hurry, these sprawling complexes of connected modular buildings can hold seven hundred to one thousand workers each. Each man gets a narrow private room with enough space for a single bed, a desk, and a dresser. A flat-screen television perches above the dresser. Meals are served in a cafeteria.

When Henderson's two weeks are up, he will head home to his wife and kids in Georgia, where he now lives. Few workers in the Bakken oil field are from North Dakota. The Bakken boom is occurring in one of the most sparsely populated parts of the country. The population of the entire western part of the state could fit in a college football stadium. The largest operator of these man camps will soon house one out of every hundred North Dakotans.

Sly Henderson's story is typical. "Everyone comes for the money. Everyone comes because they're out of work," says Byington. "The guys who make it, who stick around, come from a working background. They're used to working hard for long hours." The same could be said for the companies that are drilling wells. Byington grew up on a farm west of here. His uncle raised cattle, grain, and potatoes. When his stint is over, he gets in his pickup truck and drives ten hours west to his wife and two children in Idaho Falls, Idaho. He is considering moving them to a nearby town in North Dakota, but there are so many new oil-field workers that rents have shot up.

For Byington, the oil field was a career that began when he was nineteen years old. A few days after graduating high school, he went to work for a road paving company in Rock Springs, Wyoming, where his mother had a job. After a couple months, Byington lost the job in a round of layoffs. He put in an application at a car dealership in town and then walked into the local office of an oil-field service company. "By the time I was finished with my application," he says, "I was hired." That was twenty years ago, and he has never left the oil fields. He followed oil-field work around the western states and ended up in North Dakota. He has driven trucks and checked water levels. He spent a winter working on top of the giant blender, making sure that the liquid was mixing with the sand, while trying to keep his numbed fingers from hurting too badly. He worked in the chemical van, where the saccharine smells from giant silver boxes of chemicals bothered him. Whenever something broke, Byington wanted to take it apart and fix it so he could understand how it worked. His work ethic earned him promotions.

By the time the Bakken oil field began to boom, he had risen to the top job on a fracking crew. Byington has a calmness about him, even when subordinates get frazzled. On the Irene Kovaloff, one delay followed another. After the gasket on the pumper truck is replaced, a safety valve designed to regulate pressure in the metal pipes doesn't work. It needs to be swapped out. As time passed and daylight ebbs away, Sly Henderson's politeness starts mixing with aggravation. When his order to turn on a stand of four halogen lights isn't met with a quick response, he barks into his headset, "Gerard! Would you please go get the goddamn lights?" After a two-hour delay, the crew is ready to frack. The workers have built a factory on a promontory in an ocean of prairie grass in two or three days. After the job is done, it will be disassembled and moved a few miles to the next job.

"Blender, blender, are you ready?" Henderson calls out into his mouthpiece. "Thank you."

"Chem ad, ready to roll? Sand guys, is everything in place? Know which one you're coming out of?" A pause. "Okay. Good." The sun has set. Outside, the wind knocks a halogen floodlight from its stand, leaving it dangling by its cord twenty feet up in the air. With each gust, the light sways back and forth, and shadows dance around the well.

Henderson absentmindedly fingers the side of his computer monitor, which displays a grid of boxes six across and nine deep. The boxes record different measurements, such as pressure, pH balance, and volume. Each box has its own eye-pounding color combination: green lettering on a brown background, yellow on blue, white on pink, black on orange.

He speaks to the workers in the van. "Okay, everyone in position. Ready to go?" There are murmurs of assent.

A small, heavy black ball, less than an inch in diameter, is taken out of an orange container. It feels surprisingly heavy in the hand, like a piece of a meteor. It is measured with calipers to ensure it was the right size, and then handed to a worker who walks it outside to a set of valves near where the main four-inch steel pipeline angles off the ground and climbs a dozen feet before turning and heading down into the top of the well. The worker holds up the ball and puts it into a large opening. He screws a top onto the opening and then turns a yellow handle. The ball begins a two-minute journey to the bottom of the well, riding on a plume of thick gel. When it arrives near the end of the well, it wedges into a small baffle inside the pipe. The force of the liquid behind the ball slides open a sleeve, sealing off the rest of the well and opening several small holes. The liquid rushes through the holes and hits the Bakken, beginning the process of cracking the oil-rich shale.

Forcing liquid under extraordinary pressure into the Bakken is the raison d'être of this operation. It is why Marathon invested $9 million on the Irene Kovaloff and why men are working through the cold North Dakota night. It is why a twenty-thousand-foot hole has been drilled under the rolling hills covered with sunflower farms and cattle-grazing land. It is why thousands of men, and a few women, have migrated to this corner of the country. There is oil here—oil that will be taken to a refinery on a peninsula north of Seattle and others strung along the Texas Gulf Coast and even on the East Coast. The refineries will heat up the crude in pressurized vessels, and then break up the long strings of hydrogen and carbon molecules into smaller pieces. These pieces have recognizable names (gasoline, diesel, jet fuel, heating oil, propane) and some less famil-iar names (naphtha) that are octane additives. The crude from the Irene Kovaloff will flow through a labyrinthine industrial system: well to storage tanks to pipelines to crude rail cars to refineries to more pipelines to bulk gasoline distributors to convenience stores to car tanks.

The Irene Kovaloff is a small part of an enormous change in the United States's relationship with oil. After World War II, inexpensive oil fed an economy that roared. Affordable fuel is why America built the interstate highway system and the suburbs. Travel boomed on the back of gasoline. Route 66 wouldn't have been built, much less become iconic, if only a few could afford to drive its long stretches. The United States became an enormous consumer of oil, and also one of the world's largest oil producers. In late 1970 ten million bar-rels a day came from US wells. And then began a period of long, slow decline. But consumption didn't slow. It kept increasing. Rising imports made up the difference. Tankers full of crude arrived at the then-new Louisiana Offshore Oil Port, or LOOP, America's single largest point of entry for crude. Deepwater tankers idled a few miles off the coast and unloaded the cargos into floating buoys connected to pipelines. Within a few years, more than five million barrels a day of OPEC crude was imported into the United States. In 1973 surging fuel costs led to the first "oil shock," a period when geopolitical disputes cut off supplies and global economic growth was clipped by pricey oil. This pattern of a strong economy leading to high oil prices that, in turn, contributed to recessions repeated in 1979, 2001, and 2008. Expensive foreign oil has long been a brake on the economy. In 1973 President Richard Nixon announced Project Independence, an effort to eliminate dependence on foreign energy by 1980. Project Independence failed. By the end of the decade, imports from overseas had nearly doubled.

"I've been in this business forty-three years, and this is the biggest change in my career," observed Bill Klesse, chairman and chief executive of Valero Energy, the largest refiner in the United States. Reliance on overseas crude is plummeting, he said. Part of the reason is that cars are getting more fuel efficient and using more biofuels. A bigger part of the equation is that America and Canada are generating a lot more crude oil. Sitting in his office in San Antonio, Texas, Klesse explained that he was retooling his refineries to run on crude from Texas and North Dakota and importing less from Nigeria and Saudi Arabia. "If you said three years ago that North America could be oil self-sufficient, it was a joke," he said. "But now, it's very real."

The International Energy Agency, a Paris-based energy watchdog and forecaster funded by the world's industrialized nations, believes this shift is just beginning. It predicts that by 2020, US oil production will grow to an all-time high of 11.1 million barrels a day. Around that date, America would surpass Saudi Arabia as the world's largest oil producer. Canadian crude output also grows quickly in the forecast. As that happens, LOOP will have fewer visitors. By 2030, North America could become an oil exporter. Of course, these kinds of predictions are often wrong. But the Irene Kovaloff—and thousands of other wells like it—are at the forefront of a major geopolitical change. For generations, US foreign policy aimed to keep the spigots flowing overseas and the channels of oil delivery unimpeded by foreign potentates. As the ball traveled through the Irene Kovaloff well, speeding its way to the Bakken, all of these foreign entanglements seem less pressing.

The hydraulic heart of fracking is the liquid pumped into the well. Almost all of it is water: snowmelt from the upper Rockies that flows into the Missouri River and into the giant Lake Sakakawea Reservoir. From there, local distribution companies pump it about the state, and oil companies buy it by the millions of gallons. By the time these companies inject it into the well, it doesn't look like river water. It looks like gelatinous glop. In the Bakken and elsewhere, companies transform the water into a highly engineered viscous liquid designed to carry sand deep into the new fractures. As it heats up underground, the gel reverts to a watery state. This change allows the sand to drop out and remain in the fractures, holding them open like pillars in a coal mine. The water flows back out.

Bobby Kinsey, a fluid technician at the Irene Kovaloff, oversees this alchemic transformation. He works in a cramped lab van that looks like a mobile high school chemistry classroom. An open plastic tub of potato salad and a Dr Pepper sit near his computer. To test the frack liquid, Kinsey pours a few ounces of water into a kitchen blender attached to a large plate-sized dial that controls the speed of the blades. He turns the dial, and the water begins to churn. He picks up two small syringes with liquefied guar—a bean grown in India and used extensively by McDonald's to thicken its shakes—mixed with diesel fuel. In 2012 demand for guar rose and, coupled with a dry growing season in India, created a temporary shortage. Dave Lesar, chief executive of Halliburton, a global oil-field services company based in Houston, promised his customers that there would be no disruptions, however, because the company had created a "strategic guar reserve." Halliburton and others are developing a synthetic guar alternative so that the rainy season in India never threatens fracking again.

Halliburton has also created a frack fluid that it says contains ingredients sourced entirely from the food industry. Colorado's governor, John Hickenlooper, said he "took a swig" of it, as did Lesar, in a meeting in November 2011. "It was not terribly tasty," Hickenlooper, a former oil industry geologist turned brewpub owner, told a US Senate committee, "but I'm still alive." On its website, Halliburton warns against quaffing its "CleanStim" product.

Marathon is not using this new frack fluid for the Irene Kovaloff. In the lab van, Kinsey squeezes out a few drops into the blender and lets the mixture churn for three minutes. The liquid in the blender soon has the consistency and color of watery milk.

The next ingredient is a few drops from a plastic bottle marked "Buffer." I ask what a buffer is. He says he isn't sure. A quick search for the material safety data sheet, a federally required binder of paper that must be nearby whenever potentially hazardous substances are used, yields nothing. He digs deeper through a cabinet and finds some information. "A proprietary blend of inorganic salts," says Kinsey, a beefy thirty-year-old with close-cropped hair. He used to build custom houses in the Seattle suburbs, not mix up batches of frack fluid. The 2008 housing market collapse left him without work. He found a classified ad for Bakken jobs on Craigslist. "Washington's economy sucked, so I ended up out here," he says.

The next two ingredients are mysteriously labeled syringes

marked "30AG" and "32." Kinsey shrugs when I ask what they were. Researching them later, I discovered that the first is similar to the fuel used in camping stoves. The other is a mixture of boric acid and methanol. Kinsey puts in a few drops and lets it mix together. The liquid was gooey and pale yellow. He pours it out of the blender into a plastic cup and frowns. It is too cold, not at all like the sweltering conditions expected two miles underground, and the desired chemical reactions aren't taking place. He pops the cup into a microwave oven. After nearly a minute, the microwave dings, and he takes out the fluid and begins to pour it from one cup to the next. A tongue of the gel inches out of the first cup. He flicks his wrist expertly, and the gel jumps back into the cup. He tilts the cup, and the tongue reappears. It plops into the other cup in a large blob. Kinsey has mixed up a batch of what looks like Slime, the 1970s toy sold in small green plastic trash cans that children let ooze through their fingers.

During the frack job, workers add other chemicals to the gel. Several large vats of chemicals are in an adjacent truck. The air inside has a sweet, acrid scent. The chemicals include biocides to kill any unwanted microbes that could eat away at the gel, surfactants to make the liquid slippery so it doesn't generate too much friction on the way down the well, and inhibiting agents to prevent minerals from building up. Water and guar make up about 99.1 percent of the liquid; the chemicals are the rest. Even in such small concentrations, the volume of chemicals can add up because the Irene Kovaloff requires so much liquid. A laundry list includes 98 gallons of phosphonic acids, 118 gallons of magnesium hydroxide, and 138 gallons of 2-butoxyethanol, a compound used mostly by dry cleaners and paint manufacturers. Kinsey mixes up small batches to make sure the liquid is gelling properly. Outside his van, large spinning industrial paddles mix the chemicals, guar, and water together. It is carried through thick but flexible black pipes onto fifty-foot-long trucks. There the mixture flows into a machine that resembles a large truck engine. A crankshaft turns five plungers that suck up the fluid into a chamber and expel it. Every valve and every inch of steel are potential weak links. Pipes have color-coded bands indicating when they were last inspected. One fatigued connection, and frack fluid can end up spraying all over the pad—or an untethered, flailing pipe can kill a worker. Oil-field hands tend to be superstitious. The well head is a ten-foot-tall stack of red valves. Inside are several small brass gaskets, each with a pin-sized hole. Oil-field custom requires these holes all point north. "If they're all facing south," says Byington, "we oughta just drive off location now." He wasn't joking.

The current Bakken boom began on September 7, 2008, the day the US housing market crashed and a deep economic recession began. That day, a blue and white drilling rig broke ground on a well at noon. A couple hours before Brigham Exploration began to drill the Olson 10-15 #1H, Treasury Secretary Henry Paulson called an unusual Sunday-morning press conference to announce that the federal government was taking control of troubled mortgage giants Fannie Mae and Freddie Mac. Investment banking giant Lehman Brothers would file for bankruptcy a week later. "That was the financial crisis, and it was scary times," said Bud Brigham, the chief executive and founder of the company. Completing the well would drain precious resources. He wasn't sure if there would be any more money to keep going.

Before 2008, there were a lot of experiments attempting to frack long horizontal wells. Typically, energy companies pumped frack fluid into a mile-long section of a horizontal well and hoped it would push into a small existing crack in the rock. It was like bringing a chainsaw into a surgical suite. The technique was known in North

Dakota oil circles as a "Hail Mary frack." It didn't work. Brigham tried several of these fracks and made wells that produced just two hundred barrels of crude oil a day; not enough to justify the expense. A different company, EOG Resources, introduced swell packers to North Dakota around 2004. These rubbery membranes attach to the outside of pipes inside a well. When oil hits the packers, they swell up and seal off sections of the well. These spelled the end of the Hail Mary frack. With swell packers, EOG could slice up the five-thousand-foot horizontal leg of a well into six or eight distinct sections, allowing the frack to more effectively deliver rock-cracking pressure to a six-hundred-foot slab of rock. This technique created large, economical wells in Mountrail County, where geological conditions held large stores of oil. Turning the Bakken into a giant oil field—spanning an area only slightly smaller than California and covering parts of two states and two Canadian provinces—required more innovation.

In September 2008 Brigham Exploration gambled on the Olson well. It drilled a ten-thousand-foot horizontal leg and decided to try creating twenty distinct frack stages, slicing up the well into sections. Brigham's engineers were now trying to bring surgical tools into the operating room. But they needed to push swell packers and fracturing equipment to the very bottom of the well, including through the nearly two-mile-long horizontal lateral. No one had ever attempted to do this before. If the tools got stuck in the well, Brigham would have created an $8 million clogged pipe. The company attached metal cups to the outside of the long, slender tools and pumped in water to propel the tools to the end of the well; the concept is similar to how wind catches a boat's sails. Brigham's financially precarious situation drove it. The company ended 2008 $300 million in debt. The total value of its shares was less than $60 million. Prospects for borrowing more money weren't good, and as the well was drilled,

Wall Street essentially closed its lending window amid the growing financial collapse. The company's best shot for getting more money would be to show that it had hit on a way to drill better wells. Or more precisely, that it had found a more economical approach to producing oil.

The drilling of the well continued through the North Dakota winter and into early 2009. Temperatures dropped to 30 degrees below zero. There were four feet of snow on the ground, recalled Russ Rankin, a Brigham engineer. Motors on the drilling rig that were shut off for even a couple hours froze up and needed to be replaced. Despite the conditions, Brigham managed to force the tools into the well with few problems and fracked the well twenty times. By late January 2009, it was time to pull out all the tools and see if oil flowed from the rock.

Bud Brigham, who grew up in the West Texas oil patch but set up his company in the hills surrounding Austin, sat in his office nervously waiting for word. When the call came in, the news was better than good. It was stupendous. The Olson was flowing at a daily rate of 1,100 barrels, plus another 1.3 million cubic feet of gas. Brigham was exhilarated. "That is one of the wonderful things about the oil business," he said. "It still has the romance and the excitement. On the one hand, you have the risk of failure, and on the other hand, you have the jubilation of success."

The Olson well was easily twice as expensive as the type of Hail Mary wells Brigham had tried a few years earlier. But the well produced nearly five times more oil. In 2006 it cost Brigham about $40 for each barrel of oil it pulled from its Bakken wells. Three years later, with swell packers and precision fracks, Brigham's engineers had driven down the cost per barrel to below $16. What's more, the Hail Mary wells yielded about 110,000 barrels over their operating life. By 2009, wells using the new twenty-stage fracks—in the same places and from the same Bakken rocks—were generating about 500,000 barrels. By driving down the cost per barrel, Brigham expanded the number of wells that could be drilled profitably in the Bakken. A few years ago, five thousand Bakken wells in North Dakota alone was a dream. As Brigham and others drilled more wells using the Olson as a template, the state predicted in 2010 that there would be twenty thousand wells drilled over the next two decades. Within a couple years, it had more than doubled that estimate. Predicting the future of oil production is a tricky business. The giddiness of a boom can lead to exaggeration. Much could derail the Bakken's growth, such as falling oil prices, rising costs, and inferior rock quality as drilling expands farther out toward the edges. But regardless, the Bakken is the largest oil field found in the United States for decades.

The industry has kept innovating. Brigham added frack stages, eventually reaching forty for each well. With fracks at closer intervals, more of the Bakken's oil drained out. Other companies followed Brigham's lead. In 2011, three years after tiny Brigham flirted with financial end days, Norwegian oil giant Statoil bought it for $4.4 billion. (Bud Brigham owned about 2 percent of the outstanding shares.) By then, other oil shales had been discovered, including the giant Eagle Ford oil field in South Texas. And if fracking could unlock the oil in these areas, it could do the same around the world. Argentina, Russia, and the Middle East are all believed to have vast oil deposits in shales. In the mid-2000s, fears of "peak oil" were rife. The Olson 10-15 helped change the narrative. Crude remains a complex and constrained global market. Even if a dozen new Bakkens are discovered on the Great Plains, global oil prices are unlikely to budge much. But Brigham in the Bakken reinforced the notion that the industry could sink its drill bits into more oil than even the most dewy-eyed wildcatters had dreamed possible.

And the industry has just begun to exploit the Bakken, Brigham said. "We are still in the early innings; we are still pretty brute force at how we break up these rocks," Brigham explained. The industry estimates that it is getting no more than 10 percent of the oil in the Bakken out with its wells. (Recovery from a conventional reservoir is usually close to 50 percent.) In shales, new technologies will be developed to get an incremental 5 percent out—and then another 5 percent, he predicted. That has been the history of the oil industry, and why competitors trying to make fuels from renewable sources such as algae or agricultural waste tend to slink away in frustration. The oil industry is good at finding new ways to get oil from the ground. It musters enormous budgets and tens of thousands of engineers to the task.

Bud Brigham believes that is how it should be. "When you really analyze it objectively, oil and natural gas are just wonderful energy sources," he said. "I mean, they are portable. You can put them in tanks, you can put them in pipelines and move them from here to there. And they are cheap, really cheap. It is extremely inexpensive relative to virtually any other energy option, and it's scalable and it's portable." Exxon Mobil CEO Rex Tillerson, a Texan and Ayn Rand aficionado whose wife once gave him a first edition of *Atlas Shrugged* for an anniversary present, made a similar point in 2007, as fear of crude scarcity buoyed government support for alternative, plant-based biofuels. "All I can tell you is [that] in all likelihood, I'll get driven to my funeral in a hearse that's using gasoline or diesel," he said. He was fifty-four years old at the time.

The most compelling case I've ever heard for oil—at least for its efficient ability to deliver energy—was delivered by Stephen Chu, the Nobel Prize–winning physicist and former US secretary of energy under President Barack Obama. In a 2010 speech to a ballroom full of energy executives in Houston, he laid out his goal for the United States to deliver new, clean energy that mitigated climate change and decreased US dependence on foreign oil. He conceded that it would be tough to knock oil off its perch as the world's dominant fuel. "Why? Because oil is an ideal transportation fuel," he said. He showed a graph of energy density per unit of weight and energy density per unit of volume. The most efficient energy sources were diesel, gasoline, and human body fat. "You want to carry your energy in as compact a form as possible. Some more compact than others," he joked. Chu pointed out where the lithium-ion battery, the heart of modern hybrid and electric cars, was on the graph. To carry a comparable amount of energy, the battery required eighty times more space and weight as gasoline and diesel. His point that day was to emphasize how much work remained on new, clean sources of energy. The assembled executives, more than one thousand of them from all over the globe, smiled quietly in satisfaction. Their product was secure.

There are environmental challenges associated with fracking, as well as with burning oil and natural gas. But there's a reason why fossil fuels dominate the energy market, why crude oil powers the overwhelming majority of our cars, and why natural gas generates so much of the electricity and heat consumed in the United States. They are good, compact fuels. And with fracking, the industry has entered a new era of plentiful domestic oil and abundant natural gas. Foes of fossil fuels face a revitalized industry.

Bud Brigham founded his eponymous company in 1990 and has lived through the modern history of energy production. "Used to be we'd go drill a wildcat well, one-in-ten chance of success," he said. In the 1990s he was an early adopter of "three-dimensional seismic," an oil-field technology that allows drillers to use sound waves to find oil-bearing geologic structures underground. "When 3-D came along in the 1990s, we drilled hundreds of wells with a seventy percent to eighty percent success rate. That was unheard of, to only have twenty percent or so of the wells be dry holes. Now, with this new technology . . . we drilled over a hundred consecutive successful wells in North Dakota—these horizontal, long, lateral, high-frack-stage wells, without a dry one, all commercial, averaging 2,800 barrels a day. I mean, in our business, historically, if someone had come up to you and said you are going to drill over a hundred consecutive wells successfully and they will average 2,800 barrels a day, they would say you're crazy, that would never happen. And it has happened."

This string of success is not limited to Brigham. By the time Marathon drilled the Irene Kovaloff in October 2012, it had drilled nearly three hundred wells. Nearly every one found oil. Some were better producers than others, and some suffered from escalating costs that crimped profits. But by fracking the rock, oil flowed from the shale.

As the small black ball descends to the bottom of the well, Sly Henderson keeps a close eye on the gauges and dizzying array of numbers. Mounted on a wall is a white monitor that graphs STP-1, or surface treater pressure—a technical way of saying how much pressure is on the Bakken. The liquid is being pressed against the rock at between seven thousand and eight thousand pounds per square inch. It is more than two hundred times the pressure of air inside a car tire. Sit on the ocean floor in the Gulf of Mexico's deepest trench, and the pressure from twelve thousand feet of water above is not equal to the pressure being applied by the frack fluid. When the Bakken gives in to this assault and fractures, fluid will rush into the cracks, creating a momentary drop in pressure.

As Henderson waits for this signal, he cracks open a fresh tin of chewing tobacco. There is a small dip in pressure, but it is only a pocket of gas pushing against the heavy gel before being overcome and forced back into the rock. Afterward, for a minute, the line displaying pressure is flat, like an electrocardiogram of a patient whose heart has stopped. Then, shortly after seven thirty in the evening, comes a small dip in pressure, followed by a spike and another dip. In the span of a few seconds, the small black ball had found its way into its hole and opened the sleeve, exposing small holes through which the frack fluid could exit the pipe and fracture the rock. It looks like the patient's heart has restarted for a couple beats. The patient is alive. The Bakken has been fracked.

Outside the van, a fine patina of silica dust is blowing. A worker, his mouth covered, uses his hands to keep the sand from spilling off the conveyor belt. Crankshafts spin quickly. Chemicals and water course through dozens of pipes. The frack factory is pumping cacophonously. Inside the van, a second ball is released and a routine established that will continue for the next twenty-nine fracks. Talk turns to college football and how the University of Oklahoma Sooners and the Louisiana State University Tigers are doing. Byington says good night to everyone and heads to his pickup to drive back into town for some sleep. On the drive, he thinks about a coming break. When work on the Kovaloff well is finally finished, he plans to head west to spend time with his family. He also plans to spend a few hours in winter camouflage, hiding behind the rise of a hill, using animal calls to trick coyotes into coming to him and his shotgun.

Anthony Fish takes over as company man. A native of tiny Humphrey, Nebraska, he has worked in the oil business for fifteen years, since youthful indiscretions jeopardized his chance to play basketball for Iowa State University. The computers and the prog have taken over the job, but he remains vigilant. "If you do something wrong, drop the wrong ball, you are pretty much screwed. You are not going to have a job," he says.

I asked him if he regretted never making it to college. "The money I'm making, it's more than the kids coming out of college," he replies. "And I'm not in debt. I don't think I'll ever leave the oil fields. I tried to leave a year and a half ago but couldn't go anywhere the money was the same." According to the state, the average oil-field annual wage in 2011 was $91,400.

At a quarter to eleven the next night, the thirtieth frack of the Irene Kovaloff is completed. It takes three hours longer than expected, but otherwise the well is a success. Another crew opened up the Irene Kovaloff a couple days later. At first water flowed out. Soon came light, sweet Bakken crude mixed with the water. On October 22 the well produced eight hundred barrels of crude—a good, but not great, result. By early 2013, Marathon had pulled twenty thousand barrels of crude from the well. Considering that the oil had been locked away until the frack, it was good enough.

4

DOMINION OVER THE ROCKS

Most hail Colonel Edwin L. Drake as the father of petroleum for drilling the well that sparked the world's first oil boom, in Titusville, Pennsylvania. His success and tenacity were no guarantees of amassing a fortune. He died a pauper twenty-two years later, in 1880. When an acquaintance happened upon a destitute Drake in New York City not long before his death, he gave him a few dollars and later convinced the Pennsylvania legislature to create a $1,500 annual pension.

Edward A. L. Roberts is all but forgotten. He arrived in Titusville a few years after Drake's well and stayed. When he died in 1881, he was one of the richest men in the United States. His wealth came from a patent for a bomb, of sorts, that exploded in the bottom of oil wells. He is fracking's father.

The history of oilmen trying to create fissures in rocks goes back to the earliest years of the petroleum age. A well is only a few inches across. A wellbore could come tantalizingly close to an oil-filled seam and miss it entirely, coming up dry and bankrupting the wildcatter. But if the reach of the well could be extended, with the wellbore sending out fingers into the surrounding rock, that could make all the difference. Not long after the earliest wells were drilled in Western Pennsylvania, inventors began hawking ways to smash the rocks at the bottom of these wells. They didn't call it fracking, but that is what they were trying to do. From the early oil pioneers to later generations of petroleum engineers, a remarkable amount of brutal ingenuity was deployed to find ways to explode, pulverize, incinerate, or melt rock. For decades, men tried to break rocks with increasing ferocity and impose their will on nature. By the 1970s, there was a saying among petroleum engineers, "When everything else fails, frack it."

There is a popular notion, particularly among opponents, that fracking is a new and untested technology. There's a kernel of truth to it. Modern shale fracking began in 1998, and few shale wells are older than a decade. But researchers in oil companies were working on "hydrafracs" as far back as World War II—and they were building on earlier efforts.

The earliest insight came during the US Civil War. Roberts was a nineteenth-century tinkerer who, before he set his sights on the new oil fields, devised a popular machine to help dentists make artificial teeth. Court-martialed during the Civil War for being drunk, he was also a litigious pugilist.

Edwin Drake drilled the first commercially successful oil well in August 1859, when Roberts was still focused on dentures. New England investors recruited Drake, a railroad conductor, to go to Western Pennsylvania and see if he could find oil. He dug a well to a depth of sixty-nine feet when the drill quickly dropped another six inches. It was a Saturday afternoon, and the crew stopped work for the day. On Sunday the drilling supervisor visited the well and noticed a dark fluid floating on top of water in the pipe. He fashioned a long ladle from a piece of tin rain spouting and dipped it into the well. Up came oil. By Monday, when Drake returned, the driller had filled up several tubs and barrels with oil. A hand pump was installed on the well, which helped extract about eight to ten barrels of crude a day.

The Drake well set off an oil boom. More wells were drilled in the area, including a few small gushers that blew oil into the air. But not all wells drilled near Titusville were so fecund. The industry would later develop sophisticated technology to search for oil deposits under thousands of feet of rocks, but in the middle of the nineteenth century, luck was often the difference between a successful oilman and an abject failure. John D. Rockefeller Sr. wanted no part of this crapshoot. He built the Standard Oil empire on refining and marketing oil. He left the hapless search for crude to others.

When Edward Roberts arrived in Titusville in 1864, it was a far different place than the quiet milling town that had greeted Drake's arrival a few years earlier. It was flush with speculators, oilmen, and prostitutes. Roberts's patent for a compact machine that allowed dentists in their offices to melt rubber and create plates for fake teeth had already made him comfortable. When the Civil War began, he recruited troops for the Union and was soon appointed lieutenant colonel of the Twenty-Eighth New Jersey Infantry. By comparison, Drake's "colonel" rank was a fabrication. Before he set off for Titusville, his financial backers sent letters addressed to "Colonel E. L. Drake" to await his arrival. Local residents, impressed, welcomed him warmly.

Roberts did not distinguish himself during the war. While encamped outside Washington, DC, he showed up drunk for a dress parade. He was brought up before a court-martial in late 1862, before ever seeing a shot fired. While waiting for the tribunal's decision, the army sent his regiment to Fredericksburg, Virginia, where he participated in one of the bloodier battles of the war. Generals ordered Roberts's unit to advance through the city, across a canal, and up a hill to dislodge Confederate positions. Artillery shells rained down on Roberts and the soldiers under his command. After the battle, he submitted his resignation in "the best interest of the regiment." The army accepted it. A couple weeks later, the tribunal issued its edict. Found guilty, he was cashiered out of the military.

His wartime misadventure, however, may have been the beginning of his famed invention. Roberts said later that his idea for using explosives inside an oil well came during the bombardment at the Battle of Fredericksburg. He noticed that some of the shells landed in a small canal and detonated underwater. The weight of the water tamped down the explosion, forcing the energy sideways into the earthen walls of the canal. Roberts decided to apply this principle to Pennsylvania's new oil wells by placing an explosive device at the bottom of a water-filled well. The water forced the blast into the rocks, rather than up the well.

The Roberts petroleum torpedo was the first successful tool for fracking rocks. Roberts brought six torpedoes to Titusville. They were tin cylinders filled with several pounds of gunpowder with a percussive cap attached to the top. Roberts's hired hands lowered the torpedoes into wells by a long wire. Once in place, a piece of metal that resembled a fishing sinker slid down the line, setting off the cap and igniting the powder. In his 1866 patent, Roberts said he intended to "fracture the rock containing the oil to some distance around the wells, thus creating artificial seams, and enabling me to connect the well thereby with seams containing the oil that would not have been otherwise reached." Oil well owners were reluctant to try the Roberts torpedo, fearful that it would cause their wells to cave in. But Roberts found a taker at the Ladies' well, about a half mile north of Drake's famous discovery, on the northern bank of Oil Creek. It was a poor well that had become clogged with waxy deposits. Roberts lowered his torpedo to a depth of 463 feet and detonated it. "The explosion caused the oil and water to shoot out of the well some thirty feet into the air, and made the ground groan like a great monster in the agonies of death," noted a contemporary account. Not long afterward, the well started to flow more oil than ever before.

Roberts achieved even greater success in late 1866 in the nearby Woodin well. It had been a dry hole, never producing any oil. The first torpedo started the well flowing at a rate of twenty barrels a day. The next month, Roberts fired a second torpedo, and the well began flowing eighty barrels a day. Word spread throughout the region, and demand for his torpedoes was brisk. Soon nitroglycerin replaced the gunpowder. Within a couple years, the local newspaper marveled at the importance of Roberts's invention: "For the past three years, it has been a most successful operation, and has increased the production of oil in hundreds upon hundreds of oil wells to an extent which could hardly be overestimated. Next to the discovery of oil, no invention has done more to enrich well owners than the Roberts Torpedo."

From the beginning, some successful oilmen have tended to regard their business as a noble cause. Oil was more than a marketable commodity. Refined into kerosene, it brought light to the darkness. Roberts did not appear to subscribe to this idea. He was a capitalist and wanted to make money. He charged $200 for a medium-sized torpedo, a considerable sum at the time. His called his service well shooting.

Roberts was not the first to test explosives in Western Pennsylvania's new oil wells. But he was successful and had the foresight to patent his idea. Several lawsuits were filed against the patent, arguing that Roberts's idea wasn't so original after all. Courts ruled against these claims, and eventually Roberts obtained a lucrative monopoly. Oilmen were unhappy paying his steep rates but knew they needed to fracture the rocks to make the oil wells more profitable. A brisk business in illegal well shooting began. The scofflaws would mix their explosives during the day and then light their illicit torpedoes at night. They were called "moonlighters"—perhaps the first neologism the oil industry contributed to the English language. To stop this assault on his monopoly, Roberts hired the Pinkerton's National Detective Agency, and the "torpedo war" broke out as the hired guns ranged across the Pennsylvania countryside looking for patent breakers.

This war was also fought in the courts. Roberts—and his brother Walter, who was a partner in the company—threatened about two thousand lawsuits against oil producers who hired illegal moonlighters. By the end of the nineteenth century, Roberts was said to have filed more lawsuits than any other person in the United States. Most of the cases were settled out of court, further enriching Roberts. "By this means, the coffers of the company were filled to overflowing, and the Roberts brothers rolled up millions of dollars," wrote a nineteenth-century participant in the oil boom.

The Western Pennsylvania oil fields eventually faded, and the Roberts Torpedo dwindled in importance. Who needed to go through the dangerous, costly chore of breaking rocks once wildcatters began to find giant gushers? The first of these was Spindletop in 1901, but others arose in Texas, California, and Oklahoma. These wells were behemoths, and the oil poured out in unimaginable quantities without any explosives.

Here is the archetypal story of discovery: after days of drilling, the well begins to rumble and growl. The derrick shakes ominously. Oil rockets skyward and falls to earth as a slick rain, coating dancing, joyous, and newly rich wildcatters. Like most good yarns, there's some truth in this story. The famous Spindletop well exploded to life this way, setting off a mad scramble in swampy Southeast Texas to lease up acreage near the discovery. Wells were often so prolific that the crude was stored in man-made black ponds, but many of these short-lived gushers dried up as quickly as they roared to life, dribbling a few barrels of oil every day. When a well's pressure gave out, oil hunters went off in search of their next big find. By the middle of the twentieth century, these gushers were largely a thing of the past. In the United States, they had been hunted to near extinction. The brief era of the gusher produced fortunes for lucky wildcatters but tended to generate oil gluts that often crashed prices. Energy was cheap and plentiful. Fracking wasn't necessary.

What was left behind were a lot of wells and a lot of oil still underground, trapped in the rocks. Hollywood, understandably, focused its lens on the moment of discovery—typified by Jimmy Stewart's primal scream in 1953's *Thunder Bay* as his hard hat and face soak in black gold. But the reality of oil production, especially in the second half of the twentieth century, was much less arresting visually. It involved finding ways to free oil and gas molecules that existed inside rocks' tiny pore spaces. Not only could Hollywood not film this struggle, it was out of sight of oil-field workers. No one could see what happened a mile underground. But petroleum engineers, members of a brand-new fraternity, could try to understand and gain dominion over the rocks. It was their job to do what it took, whatever they could dream up, to smash the rocks and allow oily liquids and gas molecules to flow into the well.

Following the torpedoes, the next major advance in the

technology of wrestling oil from rocks occurred in 1932, when Dow Chemical began to use hydrochloric acid to dissolve rocks and create channels for oil. The first test took place in Midland, Michigan, near Dow's headquarters. Company engineers mixed 500 gallons of acid with arsenic, used to prevent the steel pipes from corroding. It worked, increasing the well's flow by threefold. The next year, in North Texas, another company decided to try injecting acid more forcefully into a well. This time 750 gallons of hydrochloric acid were injected, followed by oil to force the acid into the limestone formation. Before this treatment, the well had yielded only 1.5 barrels of oil a day. After the acid treatment, it flowed 125 barrels a day. A new industry was born. By 1938, some 25,000 wells had been "acidified," and individual wells sometimes were given as much as 10,000 gallons of acid. But acidization had limited usefulness. It worked in limestone but not sandstone. This limitation was problematic, since many oil and gas reservoirs are sandstone. Engineers were on the lookout for something else to open up the rocks and force wells to give up their riches.

One of the centers of the industry's research effort was Tulsa, the self-proclaimed "Oil Capital of the World." The heavyweight in town was Stanolind, which possessed girth and swagger from its origin as one of the companies formed in 1911 when the Supreme Court broke up the John D. Rockefeller's Standard Oil trust. Soon after World War II, Stanolind undertook one of the earliest efforts to gather engineers and scientists on a campus to improve the industry's knowledge and develop new techniques to maximize production. In 1952 it opened a palatial new research center on sixty acres at the cost of $4.5 million. The local newspaper marveled that it creates "a country club atmosphere, so immaculate does it keep its lawns." Only a couple of companies had independent research units, including the Texas Oil Company, later Texaco; and Standard Oil Company of New Jersey, later Exxon. The industry's early decades were full of wildcatters who relied on superstitions and luck to find oil, but after World War II, the largest oil companies wanted to rely on science and engineering instead. (Some superstitions were unknowingly rooted in geological insights. Some early prospectors drilled in church cemeteries, believing that oil accumulated underneath. Their instinct was right, although not their logic. Many cemeteries were built on hills, and these high spots were an expression of a rock fold or salt dome underneath. Later, geologists realized that these subterranean structures often contained oil and gas.)

Riley "Floyd" Farris, a studious man committed to applying science and mathematics to improve oil-field operations, emerged early on as one of Stanolind's star researchers. He graduated from the University of Oklahoma in 1935, nearly a decade before the school awarded its first bachelor degree in petroleum engineering. His interest was in cement. Then, as now, drillers deploy cement to fill the empty space in a well, locking the pipe in place. If the well is drilled through a shallow zone of water, the cement helps keep the water out of the well and anything from leaking into the aquifer. Farris wrote several papers trying to exhort manufacturers to provide better-quality cement and operators to pay more attention to how high temperatures and pressures degraded cement.

One oil-field mystery intrigued Farris. Drillers noticed that when cement entered a well, some of it occasionally disappeared. Cement wasn't cheap, and when it was lost, the well required additional costly cementing. Some wells needed up to five batches of cement before they sealed properly. (At least, that was what good practice dictated. Sometimes it was deemed good enough and work proceeded. Some states had regulations covering how to cement wells; others didn't.) The lost cement puzzled Farris. Why did some wells take more cement than the amount predicted by his slide-rule calculations? What was happening? Where was the cement going?

While this phenomenon had been noted before, Farris was the first to study it systematically. He pulled files from 115 wells and determined there was a mathematical relationship between the pressure created by the cement and the depth of these wells. His conclusion was straightforward. The weight of the cement and other liquids in the well were rupturing the rocks, creating fractures. When the cement was squeezed into the wells, some was being lost into the cracks. What if he *tried* to fracture the rock by pumping in liquid? Unlike cement, liquid could be removed after it cracked the rocks. Once the fluid was removed, he thought, perhaps more oil and gas would seep out of the rocks.

While Farris was quiet, his colleague Bob Fast was his opposite. He was friendly and outgoing, at ease both in the laboratory and in the oil field. His father worked as a tool pusher, or manager, of drilling rigs in Illinois. When oil discoveries in the Midwest petered out, the family migrated to Tulsa. In the booming city, oil-field work was easy to find. Fast received a degree in petroleum engineering at the University of Tulsa in 1943 and then spent a year fixing cracks in the wings of Douglas SBD Dauntless dive bombers during World War II. A year later, he joined Tulsa-based Stanolind Oil's brand-new research effort.

Fast was witty and had a devilish sense of humor. After many years with Stanolind (which became Amoco and later part of BP), he purchased land on the shore of the nearby Lake o' the Cherokees to indulge his love of boating. He bought a double-sailed sloop and named it the *Dammit Virginia* after his second wife. He designed the lake house himself, burying it half underground to conserve energy and, in the 1970s, covered it in solar panels to provide heat and hot water. He also installed solar panels atop his home in Tulsa. His son, Rob Fast, remembers neighbors gawking at their unusual roof, then alien to Oklahoma. But Fast loved his solar panels. "It was free energy. It was practical technology and economic," said his son.

In November 1946 Bob Fast set out to test Farris's theory in the Hugoton natural gas field in southwestern Kansas. Fast was a twenty-five-year-old looking to make a name for himself in Stanolind's research department, and he hoped that fracking the Klepper #1 well would provide a wanted career boost. Colonel Roberts performed the first frack jobs using explosives, but Fast and Farris pulled off their first fracks using a liquid. Since water generates friction—requiring a lot of pumps to inject it into the well—Fast looked for a way to reduce water's friction. He needed a liquid that was slick, mixed well with water, and was readily abundant. Fast settled on napalm left over from World War II, since it was no longer needed to fuel flamethrowers and fill bombs dropped over Japan.

Worried about fire hazards, the mixing tanks and pumps were placed about 150 feet from one another, creating an odd sight in an industry where equipment was typically crowded together. The unusual well configuration caught the attention of industry spies. Numerous companies employed scouts to keep an eye on their competitors to see how deep they were drilling and if the wells were successful. In the pancake-flat area of Kansas, the spies didn't even try to hide. They often parked near a well and watched the operation, taking notes dutifully. The Chevron scout assigned to the Klepper noted the size of the pipes used and the well's depth. "Rust" was all he wrote, cryptically, under drilling remarks.

Fast pumped in one thousand gallons of napalm-thickened gasoline, followed by two thousand gallons of gasoline. He repeated this four times at different depths. He appears to have created fractures in the limestone. The pressure dropped each time, indicating that the liquid was leaving the well. When the napalm and gasoline were recovered, gas flowed out of the well. But it was about the same amount of gas that would be expected with a conventional well that had been soaked in acid. The frack job was a failure, but Fast and Farris would soon return with other experiments.

Stanolind's research into what it called "hydrafrac treatment" wasn't in the name of pure science. The company wanted to make wells more productive. By the middle of the twentieth century, the difficulty of finding large new oil fields weighed on the industry. It settled for lesser wells and focused on cutting costs and wringing every barrel out of a well. Oil companies had already invested in drilling these wells and building pipelines, so a new technology to get another 5 percent of the oil or gas in the reservoir could be quite profitable. What's more, during World War II, steel was needed to build tanks, bombers, and other military machines. A relatively shallow 4,300-foot-deep well required sixty-one tons of steel for the pipes. The oil industry was caught in a bind. It needed to increase oil production to satisfy the war effort, but it had limited supplies of steel required for new wells and pipelines. The solution was to return to oil wells that had already been drilled but were languishing. By the 1940s, the industry had drilled more than one million wells. More than half had been abandoned altogether or were producing just a trickle of oil and gas. These wells were considered played out. Fast and his colleagues at Stanolind wanted to see if fracturing wells could resuscitate old wells and make new wells more productive.

The United States ramped up oil and gas production to meet demands of fighting World War II. After the war effort wound down, demand for fuel kept rising as the economy boomed. By 1948, there were 3.3 million more cars on the road than seven years earlier. There were nearly a million more oil furnaces heating homes. With demand outpacing production, the industry was running full tilt to keep up. In February 1948 cold weather spread across the country, and fuel shortages occurred. A Chrysler plant in Detroit laid off workers because it couldn't get enough natural gas from Texas. There wasn't enough fuel oil for furnaces, and people lined up in Chicago and Saint Petersburg, Florida, to fill jerry cans. Even before this episode, the industry realized the growing urgency to deliver more oil and gas. Its search took it offshore, and in 1947 the Oklahoma City energy exploration company Kerr-McGee made the first oil discovery from a platform in the Gulf of Mexico—so far offshore that it was out of sight of land. One of Kerr-McGee's partners on the well was Stanolind Oil.

While Kerr-McGee was making history offshore, Bob Fast and Floyd Farris kept pursuing their idea that onshore wells could be fracked. The problem they faced was that while they could infer what was occurring when cement was lost, they didn't really know. They wanted to see what happened. So they drilled a nine-and-a-half-foot well into a shallow sandstone formation near their Tulsa offices and squeezed in cement. Then they dug up the well to see what had happened. What they saw confirmed Farris's hunch. The cement was fracturing the sandstone and spreading out in all directions. In one area, it had traveled more than five feet from the well. It may have spread even farther, but they had excavated only a five-foot circle. The result was so striking that they repeated the experiment a dozen more times. Fast and Farris found cement that had caused vertical as well as horizontal fractures.

As they grew comfortable with their ability to fracture rock with cement, they encountered a problem. Fractures with water would close up once the water was removed. Why not mix in some sand with the water to prop open the fractures? This was first tried in East Texas on a well that was producing less than a barrel of oil every day. A mixture of sand, crude oil, and a soap laced with metals was pumped into the well and left to sit for forty-eight hours. The soap scrubbed the oil off the rocks. When the petroleum concoction was removed, the well began to produce fifty barrels of oil a day. The engineers hoped their hydrafrac treatments would boost a well for a few weeks or even a few months. But to their surprise, the fracked wells kept flowing. It wasn't the oil industry's answer to eternal youth, but it was a revitalizing drug that kept aging wells producing like teenagers for a few years. They had widened the well's drawing radius, sucking more hydrocarbons out of the ground.

In May 1948 Farris filed a fracking patent. "This invention pertains to a method of increasing the productivity of an oil or gas well by providing lateral drainage channels in selected formations adjacent to a well," he wrote. The patent includes a description of an East Texas well. Before it was fracked, it was barely flowing. The well produced only enough oil to fill a tablespoon every eight seconds. Fast, as usual, conducted the field experiment. About 122 barrels of fluid were injected into the well: a mixture of crude oil, solvents, and aluminum soap—the latter an ingredient in napalm, now used for waterproofing. He mixed sand into the liquid, hoping that it would remain in the fractures, propping them open. The liquid was pumped in to 3,400 pounds per square inch of pressure, about the same as a top-of-the-line, commercial-grade pressure washer available today. The sandstone fractured. Liquid flowed in. Fast kept the liquid in the well for two days and then took it out. On a sustained basis, the well flowed fifty times more oil than it had before being fractured. An exclusive license was issued to HOWCO, the Halliburton Oil Well Cementing Company.

For decades, the industry had poked holes in the earth, praying to get lucky and hit a gusher. It had even found some success with brute force—setting off nitroglycerin in a well—to get out more oil. The Stanolind experiments hinted at a different future. Engineers could manipulate the earth. Deeply buried rocks could be conquered by the engineers being churned out of universities. It was a turning point for the oil industry, even if it wasn't obvious at the time. The age of the wildcatter was drawing to a close. The age of the petroleum engineer had begun. From this point on, the industry would be defined by men convinced they had the tools and science to bend rocks to their will.

In October 1948 one of Fast's colleagues wrote up Stanolind's findings and published them in the profession's leading journal, *Transactions of the American Institute of Mining Engineers*. The paper sets out the basic elements of modern fracking: pumping in liquids under pressure to create a fracture and sending in sand to prop open the cracks. The hydrafrac process boosted production in eleven out of twenty-three wells. And it wasn't expensive. "It is significant that the value of the additional oil and gas produced to date through the benefits of this process has already exceeded the combined cost of research, development and all field tests," the paper noted. The work at Stanolind was an immediate sensation. Journal editors made it their lead paper for the year.

The paper drew the attention of one of the preeminent engineers of the day, M. King Hubbert. It "was very, very important and naturally attracted a great deal of excitement," he recalled later. Oil companies adopted fracking rapidly. By 1955, less than a decade after the first experiments, more than one hundred thousand wells had been fracked.

Hubbert is best remembered today as the father of peak oil theory. His argument was that the amount of oil in the world is finite and that as production increases, it will reach a peak and then begin to decline. Drawn on a graph, his forecast resembled a bell curve. In the late 1940s, he became interested in the question of how many years of oil supply could be pumped out of the earth and set out to figure it out. At the same time, he studied hydraulic fracturing and wrote a seminal paper on the new technology. The two interests were connected. If hydraulic fracturing could significantly increase the availability of oil and gas, it would make more oil available and push back the date of "Hubbert's Peak." But he was not impressed with Stanolind's hydrafracs. In his famous 1956 paper outlining his ideas on peak oil, he noted that only about one-third of the oil in a reservoir was being recovered. The rest was out of reach. New techniques, he wrote, "are gradually being improved so that ultimately a somewhat larger but still unknown fraction of the oil underground should be extracted."

He calculated that peak oil in the United States would occur between 1965 and 1970, and these new technologies would, at best, slow the decline on the far side of the bell curve. Despite familiarizing himself with hydraulic fracturing, Hubbert fundamentally misjudged its impact. US oil production did peak in 1970, as he predicted, and began to decline. By 2008, it was half the level of the peak. But then it started to increase again—in 2009 and each year for the next several years. We have left Hubbert's bell curve, and it's all due to the work begun by Farris and Fast.

Fracking techniques evolved quickly. By 1956, companies had begun using more and more water, with fewer additives, as a frack fluid. This ran counter to conventional wisdom, which held that water would damage the reservoir and prevent oil and gas from escaping. Laboratory engineers recommended against using water, but early efforts in the oil fields proved successful. Cheaper than crude oil, water allowed companies to increase the amount of fluid they injected into wells. Fracks that used five to ten times as much fluid as the early Stanolind efforts became commonplace. Injection rates increased twentyfold, pumping in more fluid to put more pressure on the rocks and create more fractures. In addition, innovative pumping equipment added more horsepower to the job.

Bob Fast and his colleagues at Stanolind continued their work on hydraulic fracturing through the years. The company grew interested in using highly explosive rocket fuel to make larger fractures. This decision was ill fated. On November 11, 1970, a work crew drilled a hole to test the fuel as a frack fluid. A piece of equipment was backed into an electric line, somehow triggering an explosion. The blast killed eight workers and blasted a hole five feet deep and twenty-five feet across. Fast, typically the on-site supervisor, wasn't there. He was away on his annual deer hunting trip. His son said he had survivor's guilt and wondered if he could have prevented the accident had he been there. Fast retired a couple years later with an impressive list of accomplishments: thirty-five patents and twenty-five published technical papers.

Hubbert's famous paper isn't all doom and gloom. He was a pessimist about the longevity of oil supply, but he believed that coal's abundance would provide fuel well into the future. The possibility of nuclear energy excited him. When he presented his paper, the world's first commercial nuclear power plant was under construction in Seascale, England. Bob Fast shared Hubbert's enthusiasm for nuclear energy. Years later, Fast would get mad at the television whenever the news showed antinuclear protests. He grew up in the Great Depression and was unable to join the Boy Scouts because his family didn't have enough money for dues. He drew a line connecting abundant energy and the postwar American boom. To him, protesting nuclear energy was tantamount to protesting cheap energy and economic well-being.

By 1959, the oil industry was also interested in the power unleashed by nuclear reactions, but for an entirely different reason. It
wanted to use nuclear bombs to frack wells. Edward Teller, a father
of the hydrogen bomb, convened a meeting that year at the Lawrence Radiation Laboratory—now the Lawrence Berkeley National
Laboratory—to discuss peaceful uses of nuclear power. Teller suggested it could be used for mining and excavation. The US Atomic
Energy Commission agreed and created Project Plowshare, named
after the biblical verse from the book of Isaiah: "And they shall beat
their swords into plowshares, and their spears into pruning hooks."
The program focused first on using the power of the atom as a massive earthmover. The government toyed with the idea of using bombs
to carve out a new deepwater harbor in Alaska and build a new canal
through Panama. None of these ideas ever made it off the drawing
board, beset by technical problems and environmental worries. But
a partnership between the government and El Paso Natural Gas
became a reality. The Plowshare scientists wanted to know whether
using nuclear blasts to fracture rocks around wells would work and
be cost effective. "Aspects outside the scope of a technical program—
political, sociological, and psychological considerations—were not
matters of AEC [Atomic Energy Commission] concern," notes *The
Nuclear Impact: A Case Study of the Plowshare Program to Produce
Gas by Underground Stimulation in the Rocky Mountains*, a 1976
book about the program written by Frank Kreith and Catherine B.
Wrenn. This oversight would doom nuclear fracturing, as would
another problem.

In 1967, scientists detonated a twenty-nine-kiloton bomb outside of Farmington, New Mexico. Cheered by local civic leaders
and state officials, the bomb was lowered three-quarters of a mile
into a gas well, nestled in a shale rock formation. The resulting blast
cleared out a cavity about 160 feet across. Called Project Gasbuggy,

the blast worked. But the gas that flowed into the well contained radioactive tritium and other isotopes. Plowshare scientists decided to get bigger, in an effort to get out more gas and improve the economic return of spending tens of millions of dollars on bombs. The next detonation was called Rulison, after a town on the interstate in western Colorado. The Rulison bomb was bigger—forty-three kilotons—and exploded deeper in the ground. (For comparison, the bomb dropped on Hiroshima, Japan, in 1945 was about thirteen kilotons.) With little fanfare, the Rulison bomb was detonated in September 1969, in the waning days of the summer of Woodstock. This time measurements indicated rock fractured 250 feet from ground zero. It was called a "rubble chimney," a term to describe the extent of the smashed rock. When gas came out of the well, it was commingled with high levels of tritium and krypton-85. The Atomic Energy Commission studied potential exposure if the gas were put into pipelines and delivered to homes. The two cities that would receive the highest dosage of radioactivity from burning the gas to heat homes or cook food were Rifle, near the bomb site, and Aspen. The dosage was small, but this was only a single well.

These muscular attempts to smash open the rock caught the attention of the White House. In a 1971 energy speech, Richard Nixon talked about how finding more natural gas will "be one of our most urgent energy needs in the next few years." And he threw his support for "nuclear stimulation experiments which seek to produce natural gas from tight geologic formations which cannot presently be utilized." With backing from the top, Project Plowshare and its industry allies upped the ante. The next test would explode three nuclear bombs simultaneously—each one larger than the bomb used for Gasbuggy. They would be placed far enough from each other that the impact zones would create a vast vertical column of cracked rock from which gas could flow.

Advocates believed that they could blast their way out of the energy deficit that the United States was entering. They hoped it could become a common oil-field tool—"something we can use any day of the week in any gas field," said an El Paso engineer. The Rio Blanco blast, also in western Colorado, took place in May 1973, at a time of natural gas shortages. A couple months earlier, the reality of the energy crisis hit home when Denver public schools had shuttered because there wasn't enough gas to heat the buildings. Project sponsors of the blast created a newsletter—the *Rio Blanco News*—and announced in the first issue, "Gas from Project Rio Blanco Unit Could Equal 10-Year Supply for the States." This optimism did not match the results. The three bombs created rubble chimneys, but the fractured rock from each blast didn't connect. Gas emerged only from the uppermost blast. Instead of a ten-year supply of gas, the main legacy of the blast is an official plaque at ground zero warning against digging the soil or drilling down without permission from the government.

Undaunted, Project Plowshare planners kept getting more ambitious. The next test, called Project Wagon Wheel, involved five hundred-kiloton devices. And this array was just the beginning. If successful, the Atomic Energy Commission and El Paso envisioned forty to fifty detonations a year in southern Wyoming. But this time the atomic promoters and engineers met their match. Local residents organized to stop it. There was concern about the impact of so much earth shaking on local roads and irrigation ditches. The economics of nuking gas wells was also questionable. The Department of Energy later pointed out that $82 million had been spent on the project, but even if the gas wells flowed for twenty-five years, only a fraction of the cost would be recovered.

It is not clear when Wagon Wheel was killed, but Congressman Teno Roncalio, a Democrat who was Wyoming's only member of the

House of Representatives, delivered the coup de grace. In January 1973 he was appointed to Congress's Joint Committee on Atomic Energy. A week later, Roncalio announced that funding for Wagon Wheel was being removed from the federal budget. In 1978 Roncalio decided not to run for reelection. An opening emerged for an aspiring young Republican. He would win the seat and go on to play a crucial role in the spread of hydraulic fracturing as chief executive of Halliburton. His name was Dick Cheney.

While interest in nuclear fracking disappeared, concerns about energy supply grew stronger. In November 1973 Richard Nixon pledged to end oil imports by 1980. He failed, although he resigned less than a year after making this vow. In a short speech on energy, he asked Americans to turn down their thermostats by six degrees. "My doctor tells me that in a temperature of sixty-six to sixty-eight degrees, you are really more healthy than when it is seventy-five to seventy-eight, if that is any comfort," he said. It wasn't. Natural gas was in such short supply that Congress passed a law in 1978 that essentially outlawed the construction of new gas-fired power plants. By the time the law was repealed nine years later, the United States had built 81 gigawatts' worth of power plants that burned dirty, reliable chunks of fossilized carbon—about a quarter of all coal plants that were still in use more than thirty years later. Government officials felt they didn't have a choice. The size of new discoveries was shrinking, and four out of every five wells drilled were unsuccessful.

Faced with a full-blown energy crisis, the government responded in myriad ways. There were efforts to cut demand for energy. Speed limits were lowered to conserve fuel. And there were attempts to boost the supply of energy, including the little-known Unconventional Gas Research Program (UGR). Funding for this endeavor was fairly small: $30 million was its best year. Beginning in 1977 and for the next few years, much of the money went to a small federal research unit in Morgantown, West Virginia, allotted for the study of natural gas in shales found in Appalachia. The energy industry knew there was gas in these shales, but wells were small and unpredictable. Only in places where the shale was naturally fractured—and shallow—did energy companies bother to try. The UGR wanted to change that. It deployed geologists across the region to study rock characteristics. And it drilled a handful of wells. As far as fracking them, nothing was off the table. UGR tried fracks with chemical explosives and even freezing the rocks with cryogenics.

Located on a hill above the Monongahela River, the Morgantown Energy Technology Center began life soon after World War II as a place to research turning the region's plentiful coal into synthetic gas. Al Yost joined the center in the mid-1970s. A local kid, he grew up in West Virginia in a family that had an oil and gas business. He didn't want to leave his hilly home and wondered if new technologies could unlock the region's shales. "Conventional resources were drying up domestically, and there was a need to start looking at harder-to-get gas," he said. "We were interested in self-sufficiency, reducing our imports, and producing our domestic resources."

Over the course of a decade, Yost helped pioneer many new technologies that would set the stage for the rise of hydraulic fracturing. He and his fellow engineers placed tiny cameras in the wells to figure out what was happening and shot sound waves underground to map the fractures being created, borrowing a technology developed by federal scientists. They tried the first massive hydraulic fractures of shales—twenty years before Mitchell Energy deployed a similar approach. "Most of the industry was ignoring us or saying we don't care about these shales, we're off in a foreign country developing larger, high rate of return resources," said Yost.

One of the few in the industry who paid attention was George Mitchell. Yost collaborated with the Gas Research Institute, a private nonprofit funded by a government-sanctioned surcharge on inter-state pipelines. Mitchell sat on the institute's board of directors and helped steer its research budget and priorities. He was interested in developing new technologies to characterize the shale and create fractures. Yost and his colleagues published much of their work, some jointly conducted with the Gas Research Institute and some partnering with Mitchell Energy.

His small research allotment frustrated Yost. One large-scale fracked well pretty much drained the program's budget for a year. By 1993, the program had been shut down. The surcharge on interstate pipelines didn't fare well when pipelines were deregulated in 1992, and it was slowly phased out.

By then, however, the UGR had produced a large number of scientific papers that were printed in leading petroleum engineering journals. This blue-sky research, supported by government grants and through the Gas Research Institute, left hints of what might be possible. There was a lot of natural gas left in the United States, trapped in dense shale. There were ways to break open this rock and extract this gas. It was possible to drill wells that started off vertically and then turned slowly underground until they ran through the layers of shale. Sure, the tools needed refining and the technology hadn't gelled. But the ideas were out there.

5

WISE COUNTY

George Mitchell sat alone, poring over well logs on folded sheets of paper on the top floor of a rundown three-story building. The sun had set long ago. Outside the windows, Milam Street in downtown Houston was deserted. The room had so many filing cabinets that Mitchell worried if the floor could hold all the weight. As midnight came and went, he studied the logs, blue squiggles on thick paper, searching for secrets hidden in the lines like a Talmudic scholar might try to find messages from God.

Mitchell was determined to work harder than any other wild-catter in Houston. During the day, petroleum geologists would go to the Cambe Blueprint Library on Milam Street to search through logs and try to find overlooked signs of oil and gas. In the early 1950s, Mitchell spent his days rounding up investors and listening to deals. At six o'clock in the evening, he walked over to Cambe. The owner, John Todd, was a friend and had given Mitchell a key. He let himself in and walked up the stairs to the library. Todd had left a stack of newly acquired logs out on a table for his friend to peruse before anyone else. Mitchell worked late into the night, unfolding logs that could stretch for dozens of feet. He worked until three o'clock some nights and then would head home to sleep for a few hours.

Each log was like a single piece of a giant jigsaw puzzle. Gather enough pieces—a couple from here and a few from over there— and Mitchell hoped to grasp the picture obscured in the tessellated puzzle. Analyzing a well log is two parts science and one part artistry. Two petroleum engineers who took the same classes and attended the same lectures could look at a well log and reach opposing conclusions. One would see the potential for oil and gas; the other wouldn't. Some wildcatters had *it,* an ability to see what others didn't. One night Mitchell stared at the ammonia-based ink describing a new well drilled north of Fort Worth. He saw a hint in the squiggles that sent him to the filing cabinets, pulling out older logs from nearby wells. He thought he saw something: a layer of impermeable rocks. It showed up in the new log and also, at the same depth, in nearby wells. He unfolded more logs, gathering more pieces of the puzzle. Had he found something that others had overlooked? he wondered. Was it possible? It looked like an enormous underground container where the layers of rock had settled just so. If he was right and there was a geologic trap, then over the millennia, gas would have accumulated there underneath the stone awning. And it was just waiting to be tapped by a drilling rig.

Mitchell refolded the logs, filing some back in the cabinets and leaving the new ones on the table for John Todd to sell copies of in the morning. Mitchell had a hunch, but the only way to know if he was right was to drill. And that meant he needed to acquire mineral leases to the north of Fort Worth. He didn't have to wait long.

During the day, Mitchell and Houston's other independent oil-men gathered on the ground floor of the Niels Esperson Building, a peculiar prewar edifice topped with what looks like a stone gazebo. For those with the right combination of money, ambition, risk toler-ance, and luck, the Esperson was the first stop on the way to a mid-century career as a wildcatter. There was a bank of pay telephones at the back of the ground floor that functioned as a makeshift office for aspiring oilmen. There was a drugstore that served coffee and sand-wiches. Bar stools surrounded several zinc-topped tables. And the Esperson was one of the first buildings in swampy Houston to offer air-conditioning. Oilmen would congregate in the drugstore, plop down at tables, unfurl maps, and hash out deals.

Around the time that Mitchell saw the geologic trap during his late-night sessions at Cambe in the 1950s, a Thoroughbred racing bookie from Chicago had become a regular fixture at the Esperson drugstore. He was shopping around some leases and wanted to find a driller. Mitchell met the bookie and looked at the acreage he was offering. It was the rights to drill on a three-thousand-acre ranch in Wise County, immediately north of Fort Worth and in the neighbor-hood of his puzzle pieces. Mitchell said he was interested and struck a deal with the bookie. This deal would make Mitchell a billionaire. There was a geologic trap there. Decades later, he would figure out where the trapped gas was coming from. It was from the Barnett Shale.

Mitchell was the son of an illiterate Greek goatherd named Savvas Paraskevopoulus, who traded the Peloponnesian Peninsula for the Gulf Coast of Texas. Mitchell was born in Galveston in 1919, a seaside city that was leveled nearly twenty years earlier by one of the deadliest hurricanes in US history. The Mitchells (the name was changed by the father) lived above the laundry and shoeshine parlor run by the family. The two-story brick building is still there. It has been a grocery store, an Italian restaurant, and is now a popular gay bar. After decades in Houston, Mitchell moved back to Galveston not long after his wife of sixty-six years, Cynthia Woods, died in 2009. He lived not far away from his childhood home, in a suite in the Tremont House Hotel, a four-story nineteenth-century building in Galveston's old commercial district. He lived alone, but the staff doted on him. In the corner of the ground floor, there was a table marked Reserved, where he ate his breakfast. Across the polished marble floor, near a grand piano underneath two towering palm trees in a courtyard that reaches up to a roof full of skylights, was another small round table with wicker chairs pulled tight. It also had a Reserved sign that ensured it was available to him at all hours, every day of the year.

Businessmen and tourists checking into the hotel might have caught sight of the nonagenarian motoring about on a four-wheeled scooter, until he died in July 2013 at the age of ninety-four. A lifetime as an avid tennis player had left him with shot knees. A bumper sticker on his scooter read, "I'm an Aggies Guy," a nod to Texas A&M University, where he graduated in 1939. As a distinguished petroleum engineering student, he received a gold pocket watch from the university upon graduation. It is a tradition that continues. Every year, Mitchell would purchase two gold watches for the school. One was awarded to an outstanding senior, and the other to the most improved senior.

Over his long life, Mitchell made an enormous fortune pumping fossil fuels from the ground. He was celebrated as a visionary, an engineering genius, and the man who set off the shale revolution.

But he also lived long enough to see his legacy tarnished and his signature accomplishment criticized around the world. The jowly man who ate many of his meals at one of his reserved tables, hands dotted with age spots, holding a cane, remained a jumble of contradictions. In the midst of a successful oil career, after fathering ten children, he grew fixated on the dangers of population growth. He worried about using up global energy reserves, yet made his fortune drilling for nonrenewable natural gas. He gambled his energy fortune building a new type of green city outside Houston, but never tried to make Mitchell Energy into a particularly green company. For all his interest in sustainability, Mitchell Energy never invested in renewable energy. It was the "Mitchell paradox," said his son Todd. "He had one life spending time with global thinkers, thinking about population and resource constraints, but he never took that thread and made it a part of his business life."

Growing up in Texas before World War II, Mitchell was drawn to engineering. He was a born dreamer. One year during his childhood, he built a sled in hopes of using it in the snow. But Galveston, located on a flat sand-barrier island in the bathtub-warm Gulf of Mexico, was not an ideal place for sledding. He followed his brother Johnny to Texas A&M. During a summer break from college, the brother got Mitchell a job working in the oil patch. Hooked by the excitement, he returned to college in the fall with the goal of entering the oil business. He graduated with a degree in petroleum engineering but also took as many courses as possible in geology. It was an odd combination. Geology tends to appeal to optimists convinced there's oil just a few turns of the drill bit away. Engineers tend to be more hard-nosed realists, not dreamers. Improbably, Mitchell embodied both.

In college, he studied with Harold Vance, a midcentury giant in the trade for whom Texas A&M's petroleum engineering department is named. Years later, Mitchell recalled a bit of advice from Vance. Referring to Humble Oil and Refining, a well-known company that later became part of Exxon Mobil, the professor said, "If you want to go to work for Humble, fine, then you can drive around in a pretty good Chevrolet, but if you want to drive around in a Cadillac, you'd better go out on your own." That advice suited Mitchell's personality. He didn't want to work inside a large company. He wanted to work for himself, taking the risk and capturing the reward. A bit of a throwback to the vanishing era of Houston wildcatters, he was different in other ways from the other oilmen who gathered in the Esperson Building. His liberal streak set him apart from a city of oilmen who embraced ultraconservative causes and candidates. While his fellow oilmen built trophy mansions in Houston's exclusive tree-lined neighborhoods, Mitchell dreamed of building a new community that would become a showcase of racial harmony and pioneer the new idea of sustainable development. He eventually built the community in the 1970s and 1980s and named it the Woodlands, in honor of his wife.

By 1946, Mitchell had settled in Houston after having graduated from Texas A&M and a brief stint in the US Army Corps of Engineers. He teamed up with his brother Johnny to create an oil exploration company. Johnny was a bit of a playboy, and his charisma helped him woo financial backers. He signed up Barbara Hutton, the Woolworth heiress. Together the brothers enlisted many of the prominent Jewish families in Houston. While Johnny was the promoter, George was the geologist. He found places to drill. In 1951 he drilled on the lease bought from the bookie, which had first caught his interest in his late nights on Milam Street. The first well—the D. J. Hughes #1—struck natural gas. (Sixty years later, gas still was coming out of it.) The next ten wells also found gas. Mitchell was ecstatic. He had discovered a geological gas trap. He was convinced that this trap extended for quite some distance. Mitchell and his partners began to lease up as much of Wise County as they could afford. Within three months, they had acquired leases to nearly five hundred square miles.

Other oil companies had sniffed at Wise County in search of oil and had drilled dry holes. At least, they thought they were dry. If the geologist in Mitchell led him to see that the area had the right rocks and geological qualities, it was the engineer in him that helped him realize that getting out the oil and gas would require a new approach. Mitchell was fortunate to have acquired this acreage at the dawn of hydraulic fracturing. The D. J. Hughes well was drilled in late 1951. Bob Fast and his colleagues at Stanolind had only just begun to introduce their new hydrafrac to the world. The first paper was presented at a small gathering of engineers in Dallas in 1948. Many of the Esperson oilmen still played hunches in deciding where to drill. Mitchell made it a point to read all of the new petroleum engineering literature. This new technique intrigued him.

His backers didn't care if he used a dowsing stick or a Ouija board, as long as he had a track record of using their money to make good wells. Mitchell was convinced that using science and technology was the path to success in Wise County. As it turned out, the secret to making these wells work was to frack them. "Hydraulic fracturing had just come in about two or three years before," Mitchell said. "Without hydraulic fracturing, you couldn't make decent wells."

Using this new technique, Mitchell made good wells. He wasn't fracking into the tightly packed Barnett Shale, but into a shallower sandstone with natural fractures. By shooting in gels, they were making new cracks that connected with existing cracks, building a drainage network for the gas. And it worked, opening up the rock enough to get gas to flow. Just as Mitchell began to create large gas wells, the whole campaign almost fell apart. He needed a way to get this gas out of the Dallas market, which was small and seasonal. Another company controlled production in the area and fought Mitchell's attempts to build a pipeline that would have introduced unwanted competition. The case ended up before the Federal Power Commission in Washington, DC. It took so long to resolve that Mitchell's leases were about to expire, which would have meant losing its most valuable asset: all of the land he had acquired cheaply before word got out that there was gas underneath. Mitchell threw a $5,000 chicken barbeque in 1956 to ask the landowners to be patient. The appeal to their patience—and their stomachs—worked. Finally, approval arrived to connect Mitchell's new gas discoveries to an Amarillo-to-Chicago pipeline. Mitchell's gas would head north to keep homes warm and factories running in the Midwest.

Mitchell's discovery—called the Boonsville Bend conglomerate gas field—launched Mitchell Energy as a successful company. For decades, it was the largest discovery, by far, the company ever made. But he was scratching at the uppermost level of the gas. The real mother lode was underneath. The geological structure that Mitchell had found was trapping deep gas that over the centuries had migrated upward, seeking the lower pressure zones found closer to the surface. Something was even deeper that had cooked organic material over the millennia into natural gas. And while gas had escaped from this kitchen, an enormous amount remained inside the dense rocks of the Barnett Shale. For years after he began drilling in Wise and nearby counties, there was no real reason to give this deeper gas much thought. There was plenty of cheap-to-drill gas in the Boonsville. And there was no way to get the gas out of the shale rocks.

———

In the summer of 1973, Dennis Meadows was working at his home, a fifty-acre farm in Plainfield, New Hampshire, when his telephone rang. A year earlier, Meadows and his wife, Donella, had made the leap from obscure academics to bestselling authors. A team that Dennis Meadows had led at the Massachusetts Institute of Technology had written a computer program that forecast the Earth's future. It played with the complex relationships between food production, population, pollution, and resources, including energy. Cutting through the clutter of variables and inputs, the Meadowses asked a fairly basic question: If the Earth's population continues to grow and more demands are put on the Earth's resources, what will happen? Their conclusions were bleak. Massive economic collapse. Global epidemics. They published their work in March 1972 in a book called *The Limits to Growth*. It was, to everyone's surprise, a media sensation and became a bestseller translated into a couple dozen languages.

Limits to Growth emerged at a pessimistic time, and the book both captured and reinforced this worldview. *Time* magazine wrote a story about the Meadowses and their vision of a postapocalyptic world. "In the farm lands of the Ukraine, abandoned tractors litter the fields: there is no fuel for them. The waters of the Rhine, Nile, and Yellow rivers reek with pollutants," the article warned. Some demographers found flaws in the Meadowses' computer model, but these criticisms didn't get a fraction of the attention the book received. The message of *Limits to Growth* resonated with a public in a gloomy mood. This book's bleak vision—the human species was using up available resources—scared many people. They wanted to do something to fix the Earth's problem, but weren't sure what to do. They called the Meadowses.

It was not at all unusual for the rotary-dial wall phone in the Meadowses' house to ring with strangers calling to talk about the coming global economic collapse. The Meadowses were as generous as possible with their time, but the calls were distracting. They ran the gamut from people who wanted to donate money to someone who had a macabre solution to global overpopulation. One man claimed to have developed a virus that would wipe out a significant portion of the world's population. Fortunately, his offer of genocide didn't go anywhere. Neither did most of the offers of money.

The call that arrived in the summer of 1973 was different. When Meadows answered, an executive assistant in Texas asked him to hold for George Mitchell. After a brief pause, Mitchell got on the phone. He started by saying that he had read *Limits to Growth* and was moved. "I have a lot of kids, and I don't want my kids to grow up in a world with a lot of problems. I don't want them to grow up in a world that is collapsing," he said. "Is there anything I can do?"

The phone call began an unusual alliance between the Texas oilman and the doom-and-gloom academic. By this time Mitchell was a wealthy man. His days of deal making on the ground floor of the Esperson Building were over. Mitchell Energy was a substantial company—not a giant like Mobil or Amoco, but a respectably sized independent oil and gas explorer. Having pulled himself up from poverty to wealth, Mitchell's focus began to meander. In the early 1970s, he attended a think-tank retreat in the Rocky Mountains, where he met and fell under the sway of Buckminster Fuller, the futurist and inventor. Fuller, an iconic figure at the time, popularized the term "Spaceship Earth." The Earth's resources, he argued, were limited and needed to be used wisely, not frittered away. Fuller first spurred Mitchell's interest in growth and depletion. At the end of a few days spent with Fuller talking about global overpopulation and environmental catastrophes, the futurist asked the oilman, "What are you going to do about it?"

Mitchell's response was to pay for a conference in Houston

where futurists and thinkers gathered to discuss the challenges that faced the world. A primary focus of the conference was whether the billions of people would gobble up the Earth's resources. This was a dark, Malthusian vision of the world. Yet Mitchell, the son of a penniless immigrant whose savvy had made him a multimillionaire, found it compelling. He committed $100,000 for a prize to be given to the paper with the best idea on "alternatives to growth." The purpose of Mitchell's phone call was to convince Meadows to organize the conference.

Meadows agreed, and the first conference convened near Houston in 1975. Mitchell enjoyed being surrounded by people who talked about vanishing biodiversity and sustainable ranching, solar energy and overpopulation. "We coined the expression 'sustainability,' " Mitchell boasted later. That's not true, but perhaps he helped popularize it. Other members of Houston's Petroleum Club didn't congregate at his conferences, where Mitchell listened to academics and think tankers discuss topics such as "The Helios Strategy—A Heretical View of the Role of Solar Energy in the Future of a Small Planet." Indeed, few other business leaders attended, which bothered Mitchell. For most energy executives, then and now, growth is good. More people means more demand for energy. Economic growth also drives demand. Energy consumption increases alongside prosperity. What's wrong with prosperity? they ask. Energy provides a better quality of life, with modern hospitals, air-conditioning, and one car, at least, per household.

Mitchell disagreed. The title of the first conference, "Alternatives to Growth," reflected his way of thinking. He later donated $1 million to the National Academy of Sciences to research sustainable development, a topic that was largely neglected at the time. No one has ever totaled how much Mitchell gave over the years to support this research and conferences, but it ran to more than $10 million.

Although Mitchell hobnobbed with many leaders of the incipient environmental movement, he wasn't an environmentalist. "Environmental protection is fine, but the most important issue to me is sustainability," he said. The issue that mattered was the long-term survival of Spaceship Earth. Sustainability for him meant moving away from dirty fuels such as coal and oil. And renewable resources—wind and solar—weren't ready to replace them. That left one fuel that was cleaner and available: natural gas. And he was ready to supply it. He had a gut feeling, a geological hunch, that his leases in North Texas held a lot more gas than anyone else thought.

One night in 1977, Darwin K. White looked out the kitchen window of his house, a couple dozen miles northwest of Fort Worth. A yellow glow from his well house caught his attention. At first it didn't register. Why was the small shed that housed his water well glowing? Then it clicked, and he raced outside. By the time he crossed his backyard and threw open the door, flames were coming out of his well. The shed walls were scorched. His three-hundred-foot-deep water well into the Trinity Aquifer, which he used for drinking and raising a few head of cattle, had produced excellent water since he had moved in three years ago. Now it was pumping up natural gas.

"We felt that some way the gas drilling must be responsible for it," White told me years later. "Less than a quarter mile to the west, a fellow had a well on his place. I thought his well might be related to our water problems. He didn't take to that very well. His gas well was the best thing that ever happened to him, and he couldn't believe it could be contaminating anyone's water. I didn't have the resources or energy to pursue it, but I know there is gas in the well because I saw it burning."

White, who was then a hydraulics engineer at the Lockheed

plant in nearby Fort Worth, lives in southern Wise County. A couple months before the fire in his well house, a local utility district began to lay pipes to provide municipal water to his neighborhood. He got a connection to the new public system, but losing a good water well like that didn't sit right with him. He wrote a letter to the state's oil and gas regulators. And then forgot about it. He doesn't remember ever hearing anything back. The letter, however, prompted a state investigation and the first serious attempt to look at the natural gas wells that had made George Mitchell wealthy. By the end of 1977, Texas officials had investigated White's well fire and other instances of wells spouting gas. Gas was showing up in the Trinity Aquifer in southern Wise County, the state concluded, and the "probable source" were wells that had inadequate surface casing. Mitchell Energy had drilled the majority of the wells in the area.

When wells are drilled, the uppermost section needs to be encased in cement. The cement extends downward, sometimes hundreds of feet, creating a barrier between the well and the shallow water aquifers. But as state officials examined how deep Mitchell's surface casing was—and where the top of the Trinity Aquifer was—they realized that the surface casing wasn't deep enough. All there was between the aquifer and a gas well was a steel pipe. If the steel pipe developed a hole—and steel left underground in the hot earth, exposed to corrosive gas and liquids, will develop a hole—there was nothing to prevent gas from getting into the water reservoir. The Texas Railroad Commission, the oddly named state oil and gas regulator, decided to take enforcement action.

Keeping oil and water from mixing might seem like a modern concern, but it dates back over a century. Back then, the issue was keeping the water out of the oil. Today it is keeping oil—and gas—out of the water. The first efforts to improve casing were in California, where a number of large oil fields were discovered around the turn of the twentieth century, including several around Bakersfield, north of Los Angeles. New wells began pumping increasing amounts of water. The water was fresh and was suspected to be entering the oil reservoir from abandoned wells that weren't plugged adequately. In 1907 the Kern Trading and Oil Company hired five graduates of the Stanford University Department of Geology and Mining to study the subsurface geology of the new oil fields. They mapped the depths at which wells encountered water. The company didn't want to protect the water. It wanted to protect the oil reservoirs. Too much water, after all, could wreak havoc with the oil reservoir and turn a productive field into one in which gallons of water needed to be pumped out to get a teaspoon of oil. Some companies realized it was in their interest to spend extra money to protect the long-term viability of their wells. Others weren't so enlightened.

By 1915, the value of protecting the new oil industry from careless operators was clear to California officials. A few sloppy operators could gunk up an entire oil field. To protect the oil reservoir, all wells needed to be cased adequately. The state legislature passed laws to require oil companies to set surface casing to isolate water from oil. The California State Mining Bureau created a division of oil and gas operations "for the purpose of supervising oil-field operations, with special reference to the matter of shutting off water in the oil fields and conserving the state's oil and gas resources." Oil companies were assessed a fee to pay for the new division. The rules required that "casing be set and cemented and, after a prescribed interval, be tested in the presence of a representative of the division to prove that all upper waters were definitely excluded." Failure to do so resulted in a long delay in obtaining a permit to drill.

Royal Dutch Shell was the first in California to organize its subsurface geologists into a new working group. Other companies followed. Names for these new workers varied: exploitation engineer, petroleum production engineer, drilling engineer, and production engineer. Eventually the industry settled on petroleum engineer. Universities started to offer classes to train people to work in the oil fields, figuring out how deep to set the surface casing. Engineers began to study the reservoirs themselves, to determine where to drill wells to get as much oil and gas out of the ground as possible. The University of California at Berkeley and Stanford were among the first schools to offer classes. The new discipline spread to the University of Tulsa, West Virginia University, and the Missouri School of Mines and Metallurgy. The University of Pittsburgh conferred the first petroleum engineering degrees in 1915. Four students initially selected the degree; then one switched over to become a mining engineer. The three remaining recipients are notable because two of them were Chinese. One is listed as coming from what is now called Guangdong Province, and the other from Sichuan Province. The third hailed from New Castle, Pennsylvania.

Several decades later, by the time Texas investigators were looking into Mitchell's wells near Darwin White's house, the importance of keeping water separate from oil and gas was well established. State investigators proposed that Mitchell and every other company with a gas well in southern Wise County should add more cement. Mitchell Energy did not like the proposal at all. If there is contamination, its executives responded, the problem was probably some old wells that were leaking. The state had required surface casing down to 300 feet until a few years earlier. If the state wanted aquifers protected down to 450 feet, that wasn't Mitchell's problem. It had followed the rules.

Government investigators held their ground. The wells needed cement down to 450 feet. They proposed firing holes in the pipes and squeezing in new cement that would be forced upward, creating a new barrier keeping the gas and water apart. Mitchell Energy executives argued against this suggestion strenuously. The proposal, covering only a small section of the entire Boonsville Bend field, would require remedial cementing of seventy wells. Even if fixing every well went smoothly, it would cost $665,000, they argued. Moreover, at the end of the well's useful life, Mitchell couldn't fish out the steel pipe and resell it. Another $560,000 loss. "If the Commission proposal was later expanded to include the entire Boonsville field, Mitchell would be obligated to a loss of several million dollars!" Mitchell officials argued in a written response. "This is too severe when considering that the operators in this field would be penalized for following the very rules set up by the Commission." Mitchell proposed checking on its wells weekly. If gas was leaking behind the pipes, it would see a buildup of pressure at the surface. The state relented. Mitchell wasn't required to pay a fine or fix its wells.

Darwin White still lives in the same house. He was unaware there had ever been an investigation by the state. "I thought it went to naught. We didn't ever pursue it. You can't tell what is going on underground, and we couldn't get anyone to investigate it," he said. He still gets his water from a local public water system. His 1977 letter to the state was the first of many complaints. Other accounts of gassy water wells followed. Eventually a water treatment plant was built on the western edge of the county to provide municipal water from a nearby lake. The system would later be expanded to serve an even larger area. In 1997 a seventy-mile pipeline was built to move lake water to more of Wise County. Mitchell Energy, to settle a lawsuit, paid for most of it.

Mitchell acquired so much acreage in Wise County that for two decades his company kept drilling and growing. But by the 1980s, Mitchell began to wonder how much more gas was left in the Boonsville Bend field—and whether it was enough to fulfill his contractual obligation to deliver one hundred million cubic feet daily of gas into the Chicago-bound pipeline. The giant geological trap he had found was being depleted by thousands of wells. He needed to find more gas to feed into the pipeline. For years this contract had been a godsend. It offered good prices for the gas, keeping Mitchell Energy profitable and comfortable. But the contract was starting to look like a millstone around the company's neck. Contractually obliged to supply gas for several more years, Mitchell wasn't clear where that gas would come from. If his company couldn't find enough gas, it would have to buy the gas itself.

Mitchell Energy began looking at buying some wells and acreage held by other companies in the general vicinity. One possibility was a well drilled by tiny Argonaut Energy. It was 7,700 feet deep, about a thousand feet below the rocks where the natural gas in the area was typically found. As Argonaut had drilled deeper, its drill bits had also traveled backward in time. The geologic trap that Mitchell had discovered in the 1950s was deposited millions of years ago. As Argonaut went farther into the earth, it had penetrated rocks deposited even earlier. Argonaut had found the source rock. Over millions of years, leaking gas had migrated upward until it encountered the geologic trap that had made Mitchell so excited and so rich. Argonaut ran a gas detector. As the drill bit churned through a layer of shale rock, the device's readout jumped, but no gas flowed out. Argonaut plugged the well as a dry hole. Mitchell's people, reviewing the data, recognized that a similar 250-foot wedge of shale existed a few dozen miles away in a well it had drilled. It too had registered gas.

In 1978 the federal government set price controls on natural gas. Gas was in short supply, and the government wanted to encourage more exploration. Gas from existing wells received the lowest price. But gas from new wells got a higher price, and gas from wells in "unconventional" reservoirs that required fracking got a much higher price. In 1982 Mitchell Energy applied to the state for designation of a new field that would qualify for the highest prices. The new field was named after a town in Wise County. It was called the Newark East (Barnett Shale) gas field.

In June 1982, at Mitchell's personal insistence, the company fracked the Barnett Shale for the first time. A few years earlier, a government-funded project had tried a massive water frack in Wise County, right in Mitchell's backyard. The experiment mixed water and crude oil into a gelled emulsion called a "Super K-Frac." The pressure had ruptured the piping, leaving a hole about a mile deep. It took a month to locate and cement the hole. "It appears that this technique is probably not economically feasible," noted a government report on the effort. Mitchell opted to use 1.5 million cubic feet of nitrogen to frack his well. Then in 1983 he tried again with foamed carbon dioxide and water. The result was that the shale flowed at 240,000 cubic feet a day. Considering all the time and expense, it was a terrible well and wasn't profitable. (A good modern Barnett well can begin flowing at 5 million cubic feet a day.)

But that didn't stop Mitchell. The company had managed to frack the Barnett Shale, and gas had dribbled out. Mitchell directed a study of how much gas was in the Barnett Shale. The answer was billions or possibly trillions of cubic feet. "It was a substantial amount of gas if we could break it away from the rock," he recalled. With government price support and the driving need to produce more gas, Mitchell unleashed his engineers on the Barnett Shale. For the next few years, Mitchell deepened about ten wells a year into the Barnett Shale, testing different types and sizes of fracks. He was persistent because, in his view, the company's survival was on the line. If an employee balked and said he wasn't sure if it was possible to frack the Barnett Shale, he was told that this type of thinking would lead to a new job at a different company. Many people thought the project was a money sink, but Mitchell owned more than 50 percent of the stock in the company. His staff could disagree, but it was pointless.

Slide rules were still commonplace, but by 1982, every company engineer received an IBM desktop computer. Mitchell didn't want wells drilled on hunches. Data were collected and interpreted. Mitchell Energy started crunching numbers for insight and advantage a decade before this became normal practice in many industries. Wearing shirtsleeves and a tie—Mitchell's seldom-worn sports coats, one former executive said, made him look like a used-car salesman—he would check in with his engineering and geological staff on Mondays. He wanted to know about what wells had been drilled over the past week. What did the logs run to get a better sense of the rocks look like? If he heard something that interested him, he would go right to the source. It was not at all unusual for midlevel engineers to pick up the phone and hear the boss's voice bombarding them with questions about a new well. One weekend, a well was drilled, determined to be unsuccessful, and plugged. When Mitchell heard about it on Monday, he ordered that the plug be drilled out before the rig moved on. The well could come in handy, he said.

"He has a mind that people often refer to as persistent," said his son, Todd. "To me it is different than persistence. It was a form of obsession. He has a theme, and he would stick with it and stick with it." And during the 1980s and into the 1990s, he was convinced that the Barnett Shale held an enormous volume of gas. The only question was whether the company could develop an economical way to get it out.

In 1964 Randy Miller's family moved to Glenpool, Oklahoma, a city that shares a name with one of the largest oil discoveries ever made in the United States. He was thirteen years old and lived on a hog farm in the middle of the giant Glenn Pool oil field. Farmers drove their tractors around leftover oil well detritus.

A year after the move, he worked in the oil field for the first time. His job was to haul beat-up drill bits to a forge where they were reheated and reshaped. Old cable-tool rigs were in use, and the repeated dropping of the bits onto the rock would deform them. These cable-tool bits are now obsolete, replaced by bits that turn and chew up the rock. As Miller got older, he graduated to working with sledgehammers, pounding the bits after they were heated. By the time he got to college, he had started driving trucks. It was a job he would keep for eight years as he worked his way through college and then law school.

Miller's background helped him connect with jurors. He came across as a regular guy, with a theatrical flair. He once opened a Mason jar filled with hydrogen sulfide to give the courtroom a whiff of the noxious rotten-egg smell to which his clients had been exposed. A juror vomited, and the court had to recess, with the courtroom windows opened to clear out the air.

In 1994 Miller joined a lawsuit against Mitchell Energy. Not long before, Carrie Baran, a Wise County resident, had called up Gardere & Wynne, a sizable Texas-based law firm that typically represented companies charged with oil-field pollution. Her call made its way to a firm attorney named Bill Keffer. A former in-house lawyer for oil giant Atlantic Richfield, commonly known as Arco, who would later serve as a Republican member of the state legislature, Keffer was intrigued by what he heard from Baran. Gas in her water. Rotten-egg smells. Keffer roped in Miller, and together they started looking into the matter and gathering more plaintiffs.

They spent a year looking at well records and found that many of Mitchell's wells were "short cased." Mitchell hadn't spent the time or money to make sure the aquifer was protected, they later contended in court. The protection—cement and extra layers of pipe—didn't extend below the base of the aquifer. The Texas Railroad Commission had made the same finding in 1977. Miller spent a year gathering more information about Mitchell Energy in Wise County. The most damning evidence was what he came to call "Post-it gate." Years later, as Miller explained how it worked, his speech grew impassioned and voluble. He moved his head to punctuate his points, his wavy white hair—matched with a cropped white goatee—bouncing along. He had ordered up government documents and received boxes and boxes of photocopies. After spending hundreds of hours reviewing them, he noticed something odd. Someone had affixed pink Post-it notes to the records that had been photocopied along with the rest of the document.

The notes provided evidence of overwhelming mismanagement by the state. When a request was made to drill a well, the state's water board was consulted and reported back the approximate depth of the freshwater aquifer. If the aquifer is 500 feet deep, the energy company needs to make sure it runs surface casing down to 550 feet. At least, that was how it was supposed to work. On the official documents, the state oil regulators had okayed wells where the casing was deeper than the freshwater. Then a pink Post-it note was affixed showing the true depth of the freshwater. Well after well had been approved without enough protection for the water.

Before the trial, Miller met with Mitchell's attorney, who offered $70,000 to settle the matter without a trial. Miller balked. What followed was an epic legal fight that spanned two courtroom encounters and went all the way to the Texas Supreme Court, leaving both Miller and Mitchell scarred.

Miller had a massive health scare in the middle of one of the trials. During a weekend break, he went home to Tulsa and collapsed while driving his car, with his thirteen-year-old son in the passenger seat. The car careened off the road, but neither Miller nor his son was badly hurt. Miller's doctors, however, said he was perilously close to a heart attack and ordered him to take himself out of any stressful situation. A new lawyer was brought in to finish the trial.

Mitchell also suffered. He was accused of exploiting the people of Wise County, extracting millions of dollars' worth of gas, and leaving behind contaminated water. In 1996 a jury handed up a record-setting $204 million verdict—$200 million for punitive damages—against him for fraudulently concealing information about leaking wells from state regulators. The verdict was extraordinary. It was more than the company had made in the previous four years combined. Carolyn Etheridge said that her fellow jurors felt the company didn't care about the community and just wanted to make as much money as possible. She said the jury was angry and thought a large fine would get Mitchell's attention. "We wanted it [the pollution] stopped. It was a small sum of money, in our minds, to fix the problem."

Mitchell was flabbergasted and furious. He told a reporter at the time that Wise County was a "burned-up, parched, miserable place" before he arrived. His leases and gas development had brought wealth to the county. He hired a raft of top Texas lawyers to handle the appeal. The verdict was so large that it threatened Mitchell Energy's existence. Mitchell had to sell his beloved Woodlands real estate development to plug the gaping hole in the company's balance sheet created by the verdict. "I have never believed, nor do I believe now, that Mitchell Energy Corporation is the cause of the problems that the plaintiffs are complaining about," he said at the time. Years later, he blamed Miller for hoodwinking the jury. "Smart lawyers would convince a jury of anything. That's what happened. And then we got some of the best attorneys in the country; we had nine of them, at four hundred fifty dollars an hour for three years, to fight it. And we got all the data together, and we beat the hell out of them," he said.

Miller saw it differently. To Mitchell, Wise County "was a cash register and nothing else," he said years later in an interview. "It wasn't his home. It was someone else's home. It was where he got his wealth from, but it wasn't where he had to live. He built his wealth there. He had done it in that Wild West manner of rotted casing, falsified records, drilling, and cheating the locals and politicking administrative agencies—he had played that rough old game, and by the time it had got out of hand, he wasn't about to own up to it. Neither was that company. We made them own up to it in front of that jury."

In November 1997 an appeals court reversed the lower-court ruling, concluding that the statute of limitations on bringing the case had expired and that the landowners' experts didn't present conclusive evidence that Mitchell was responsible for the gas in the Trinity Aquifer. The higher court took the uncommon step of rendering a new verdict. Mitchell owed nothing to the landowners. It erased the $204 million verdict, a decision later upheld by the Texas Supreme Court.

The Trinity is one of the largest underground bodies of water in Texas and one of the most important. A century ago, wells drilled into the Trinity supplied fast-flowing water. An entrepreneur in Waco, Texas, created the Artesian Manufacturing and Bottling Company to sell Trinity water. He soon realized that adding sugar and coloring to the water made better business sense and created Dr Pepper soda. But decades of drilling drew down the Trinity. "Water used to squirt out of the ground in Dallas and Fort Worth. It is now 1,110 feet below land surface," said Robert E. Mace, a state hydrogeologist. The deepest drawdown occurred in and around Fort Worth, including Wise County.

According to Mace, as the Trinity emptied, the underground map was completely redrawn. Water that had moved through the rocks in one direction was now heading in the opposite direction. "The aquifer is different now," he said. Pressure in the Trinity also dropped over time. There were fewer water molecules occupying the same geological container. This change created the potential for natural gas, underneath the Trinity, to get slowly sucked up into the aquifer, Mace explained. This gas had migrated up from the Barnett Shale over millions of years. When Mitchell started drilling in Wise County, his wells penetrated an aquifer already in the midst of a millennial change. And when residents started noticing that their water wells were gassy, who was to blame? Was it Mitchell and other companies drilling gas wells straight through the aquifer? Or had a few generations drawing down the Trinity and lowering the pressure enough that gas could now move upward into the depleted aquifer? Mace said he couldn't be sure.

As Mitchell's appeals moved through the courts, the publicity around the lawsuit triggered another state investigation. The results were damning. There were 112 Mitchell Energy wells in Wise and three other nearby counties that had been short cased. The protective cement wasn't deep enough to protect groundwater. In a report issued in March 1998, a Railroad Commission staff attorney stated that Mitchell Energy "deliberately misreported" the actual depth of protective casing for all but two of these wells. At least eight wells weren't plugged properly. The company was ordered to pay a $2.24 million fine to fix its wells. The fine was the largest in the history of the agency, but just 1 percent of the penalty that the jury felt was deserved.

This proposed fine wasn't enough to calm Wise County landowners, including the plaintiffs who felt that their $204 million victory had been snatched from them because the fix was in. In mid-July 1998 Governor George Bush visited Wise County, stopping first at the historic county courthouse, which had been restored with a grant from his office, and then at Decatur City Hall. Carrie Baran, whose call to lawyer Bill Keffer had launched the legal attack on Mitchell Energy, decided to confront Governor Bush at his second stop. There is no record of what she said or his response. Several days later, she wrote a letter to him apologizing for what she called her "outburst."

"I am also pro business," she wrote. "But Sir I also believe that what Mitchell Energy Corporation has done to us is despicable . . . George P. Mitchell and his vast empire has all but ruined many of us all for the almighty dollar. He has used his power and money to fight us, call us the 'Beverly Hillbillies,' intimidate and beat us. He is not even man enough to apologize for the irreparable damage his company has caused . . . I suppose the right amount of money can buy just about anything."

In November Mitchell offered to settle the state's proposed $2.24 million fine for $100,000, arguing in a letter to the Railroad Commission that its own investigation had determined that "the poor quality water was naturally occurring, but exacerbated by poor water well construction." The company letter pointed out that a jury in a second legal case had returned a verdict finding that Mitchell Energy was not responsible and that the company had found a witness who testified that he saw a water well set on fire because of natural gas in the aquifer—in the 1920s. The letter also argued that Mitchell had already spent a lot of money on Wise County water problems: $2.8 million to repair poorly cased wells; $3.6 million to install a new pipe from Lake Bridgeport to provide connections so that residents could get rid of their wells and access municipal water; and $450,000 for water treatment systems for other residents who couldn't be connected to the new pipelines. The Railroad

Commission backed down. On January 8, 1999, Mitchell sent two checks to the Railroad Commission. One check, for $100,000, was to pay a fine to settle the investigation. The other, also for $100,000, was for the state's Oil Field Cleanup Fund, used to plug wells left orphaned by companies that went out of business.

Whether Mitchell's inadequate cementing practices contributed to the gassy wells is inconclusive. Was there more gas in the Trinity Aquifer because generations of dewatering allowed gas naturally present in shallow rocks to push up into the water? Or had some of Mitchell Energy's poorly built wells created a pathway for gas from thousands of feet below to leak into the shallow water zone? Two state investigations turned up evidence of more than one hundred improperly constructed wells, and it had looked at only a fraction of 3,700 drilled by Mitchell since the 1950s. The investigation focused on a relatively small area in southern Wise County. How many more poorly cemented wells—if any—would have been discovered had the state conducted a more thorough review is unknowable. The evidence is overwhelming, however, that the state did a poor job policing Mitchell Energy.

The lawsuit had a lasting impact on the oil business. The $204 million verdict got the attention of the executive suites and prompted a new focus on cementing wells. But many oil companies also responded by supporting tort reforms to cap future damage awards. Overturning the jury award and slashing the administrative penalty also sent a message to companies that Texas still had a favorable legal and regulatory environment for drilling oil and gas wells. It set the stage for the coming boom in the Barnett Shale.

George Mitchell was happy to share what he remembered about the scene at the Esperson Building and his company's work to break open the Barnett Shale. But the sun had set on his willingness—or maybe capacity—to talk about the Wise County lawsuit. A couple blocks from his hotel, Mitchell ate dinner two or three times a week at Olympia Grill, a Greek restaurant looking over Galveston Bay. The waiters and manager greeted him by name. Even on a hot and humid day, he preferred to sit outside. Pelicans glided above the narrow water passage that separates tourist Galveston from a shipyard where giant offshore drilling rigs and platforms are being repaired and built.

One summer evening, he ordered the Greek salad and a plate of fish with French fries. His son, Todd, and I ordered moussaka, a baked eggplant dish. Toward the end of the dinner, I asked Mitchell about the Wise County lawsuit. What did it feel like to have landowners in Wise County suing him for contaminating their water? He looked up from his plate and started talking about a pipeline company that was active in the 1950s. His son cut in, trying to direct his father back to the lawsuit. Mitchell insisted that somehow the pipeline company was at fault. The answer didn't make any sense. I looked beseechingly at Todd, who shrugged.

After a couple more unsuccessful attempts to get him to talk about the lawsuit, I gave up. He didn't want to answer or had deeply buried his memories of his hours-long deposition and the embarrassing verdict. And I didn't want to seem like I was berating this elderly man in front of two of his preteen grandchildren, who were also at dinner. Hoping to get a few more answers before dinner ended, I asked for his opinion on the role that government should play in regulating natural gas drilling.

"It has to play a more active role. The independents won't do it right. The government has to help keep them under control. The government needs to oversee wells for all kinds of problems," he said. Left unchecked, independent producers are too single-mindedly focused on getting the oil and gas out of the ground. "They go wild." Was Mitchell Energy one of those independents that could have used more government supervision?

Sensing that his father needed a break, Todd Mitchell shifted the conversation to his own family and the goats they raise in Colorado for milk to make artisanal cheese. "Goat herder to goat herder in three generations," he said.

I brought up the Mitchell paradox. Given his interest in sustainability, why didn't Mitchell Energy ever spend money developing renewable energy? "We put a lot of money into gas," George Mitchell responded. This answer bothered Todd, who had spent the last few years investing in renewables. "But you never put money into solar, geothermal, or wind," he said. "There are some big Houston families who have gotten interested." He mentioned the Zilkhas, a prominent Iraqi-Jewish family whose success drilling oil and gas wells in the shallow Gulf of Mexico was legendary. After selling their company for $1 billion in 1998, the family built a large wind business and then sold that to Goldman Sachs in 2005. Next, they invested in biomass, turning wood pellets into electricity. George Mitchell blinked a few times as he chewed some food and then said he wasn't impressed by renewable energy investments. "It didn't look very promising costwise," he said. "Right now it looks like solar has a long way to go."

He invested years and millions of dollars experimenting with fracturing the Barnett Shale, searching for ways to drive down the cost and make it more efficient. I asked whether he thought a similar stubborn approach to wind or solar might have also resulted in similar results. There was a long silence. George Mitchell sipped his iced tea and then speared a bit of bossa on his fork, bringing it to his mouth to chew. He was not ignoring me or answering my questions with withering silence. The question simply did not engage him.

He stared ahead. The sun was setting. Across the inlet, the sparks and intense light of welding arcs could be seen on one of the semi-submersible rigs. About thirty seconds passed in silence as Mitchell continued chewing. There was no answer coming.

6

ICE DOESN'T FREEZE ANYMORE

On June 11, 1998, Nick Steinsberger woke before dawn. He
dressed, got into his car, and headed north from his home in
suburban Fort Worth.

He was nervous. There had been layoffs recently at Mitchell
Energy, where he worked. The company had hit a rough patch.
Mitchell's moneymaker was finding and selling natural gas in North
Texas, but the company was losing money. Its cash was drying up,
and Steinsberger worried that his name would be on the list in the
next round of layoffs. He was thirty-four years old, with two kids
at home, a wife in college, and not much money saved. He drove a
1991 Toyota Camry with nearly two hundred thousand miles on it.
The car was infamous in the North Texas natural gas field where he
worked. He would pull into well sites and park next to a row of late-
model pickup trucks. As a field engineer, he was the boss at the well

site. But the foreman and everyone else drove nicer cars. His colleagues often ribbed him about his clunker.

As he drove to the well where he was working that day, the houses thinned out. One mile there would be a subdivision of new homes, the outer reaches of the sprawling Dallas–Fort Worth metroplex. The next mile it was fields with cows and bales of hay. Steinsberger was thin and energetic, with long, slender fingers. As a high school student, his fingers had helped him excel on the violin. He even performed a couple of times in an all-state orchestra in Nebraska. But unlike the truck-driving roughnecks and laborers who drilled the wells, his job didn't require much use of his hands. He designed well completions.

He took over after the well was drilled. These are not the gushers of Hollywood lore. Drilling produces a hole in the ground a mile or two deep. Once the drill bit finishes churning its way through rock, little gas flowed out of the wells he worked on. Steinsberger's job was to coax, crack, and wheedle the gas out. His tools were highly specialized. There were perforation guns: tubes inserted into the wells with explosive charges that punched holes in the rock faces that had been exposed by the drill bit. Hydrochloric acid could be sent down the well to clean up the debris left behind. And there was cross-linked gel, a heavy, goopy substance that flowed like a liquid but was engineered to have the qualities of a solid. This compound was pumped into a well under high pressure. It would flow through the holes and crack the rocks and carry grains of sand deep into the fractures. After a while, the 200-degree temperature at the bottom of the well would change the gel. The heat broke the polymer chains and made it less viscous. More of a liquid at this point, it could be retrieved from the well. The sand was left behind, propping open the newly created cracks for gas to flow into and then travel up to the surface.

At least, that was the idea. The gel didn't entirely clear out of the rock. Pumping in barrels and barrels of gel would create cracks in the shale and then plug them up. It was like building a highway with feeder roads but then leaving behind orange cones that blocked off lanes at random intervals and closed down on-ramps. Traffic could only creep along the new road. In a similar vein, some gas would make it through, but most remained at a standstill on the road. (Fifteen years later, the gel used by Marathon Oil in the Bakken was several iterations better than what was available around Fort Worth in the 1990s.)

Steinsberger had suggested a new and revolutionary idea. He wanted to use water instead of gel. And not just a little water, but a massive amount of water—four or five times as much water by volume as the typical slug of gel. It was a particularly audacious idea because he was trying to get gas out of the Barnett Shale, a dense slab of rock that was nothing like the permeable sandstones the oil industry tended to target. If the size of the typical shale pore was a marble, a sandstone's pores were the size of a lecture hall. The gas was trapped in tiny lockups without space to move around. Shale cored out of a well looks more like a piece of black plastic than rock.

As Steinsberger drove to the S. H. Griffin #4 well that morning, his new approach to fracturing wells was not going well. The previous year, he had pitched the idea to his bosses at Mitchell Energy headquarters in the Woodlands. He had received permission to use water to fracture three wells, but not without resistance. "A couple of the managers thought I was an idiot for trying," he said. One told him that he would eat his diploma if the idea worked. How would water break open impenetrable shale?

Over the course of several months, his first three wells had failed. Sand had built up in the wells and clogged up the perforations. Petroleum engineers call this phenomenon a "screen out." Steinsberger went back to the Woodlands and asked for permission for three more wells. A fourth well in March had also screened out. The S. H. Griffin #4 was his fifth. His bosses' patience was running out. Using water to crack shale was an outlandish idea. "It is counter to everything you were taught in school," said Ray Walker, an engineer at another gas company who was friendly with Steinsberger. "It was contrary to everything we had all been taught about fracturing." Steinsberger knew he was putting his career at risk and worried that more failure would hasten the pink slip he believed was in his future. He wanted to make a name for himself inside Mitchell Energy and had begged his bosses for a chance to try this new way of fracking.

Real-world challenges drew Steinsberger into engineering. His family moved around a lot when he was a child, going where his father landed jobs as a political science professor. He grew up in small towns in Wisconsin, Missouri, and Nebraska. His father remembers that his son wasn't particularly mechanical, just curious and independent. "He wouldn't argue or fight with you, just say okay, and then he just did whatever he wanted to do," said George Steinsberger. His father encouraged Nick and his sister to stay engaged with politics and encouraged them to read. On a summer vacation when Nick was about fourteen, the family drove west across the country. George remembered his son reading Joseph Heller's *Catch-22* and laughing out loud at the satirical novel about bomber pilots.

Young Nick was also drawn to articles about what people built. One article that caught his attention was about the THUMS Project. Named after the five companies involved—Texaco, Humble, Union, Mobil, and Shell—THUMS was a clever solution to a daunting engineering problem. The giant Wilmington oil field underlies Long Beach, California. It is one of the five largest ever discovered in the United States and has been producing oil continuously since 1932. More than two billion barrels of oil have been extracted from loose sands that run for miles along the coast, stretching under the city and into the harbor. As the oil was taken out, concern grew that the weight of the earth above could settle, lowering the city and causing problems with foundations, streets, and pipelines. The oil field's operators began to pump water into the sands to replace the oil they were taking out. Where once Long Beach floated on a sea of oil, engineering ingenuity substituted a sea of salty water. "I wrote a sixth or seventh grade paper on the THUMS Project," Steinsberger told me. "I thought that was about the coolest thing in the world." That set him on the course to become an engineer.

By seven o'clock on that June day, the sun was rising as Steinsberger drove up Tim Donald Road to the well. On one side of the road was a row of double-wide prefab houses. On the other side was an open field of scraggly brush. When he pulled off the road onto a dirt pad, the engineering muscle assembled there was a sight. There were twenty water tanks, each the size of an eighteen-wheeler, and more than a dozen pump trucks with the powerful engines to force the water into the rock. There were a couple trucks that supplied chemicals to be mixed into the water. Among the chemicals was a friction reducer to make the water more slippery, as well as a biocide to kill any organisms that were stowaways on the nearly two-mile journey from the surface to the end of the well. There were trucks filled with sand and blenders to mix together the watery cocktail. All in all, about forty people were at the well. The assembled workers knew that this frack job was different. Even if they hadn't been briefed on Steinsberger's water idea, there were twice as many water trucks and pumps as there were on a typical job. The final vehicle on the scene was Steinsberger's beat-up Camry.

He got out and talked to Mitchell Energy's foreman to make sure the preparations had gone smoothly. Steinsberger walked around the well pad, checking the connections of the aluminum irrigation pipes that crisscrossed the ground. Then he went into the command center, a trailer rigged with computers and monitors. He went over how he wanted the job to proceed one last time and then gave his assent. "Let's go," he said.

Workers in the trailer began firing up engines. The first of more than one million gallons of water flowed deep into the earth. It was enough to cover the drilling pad up to Steinsberger's chest. Inside the trailer, Steinsberger watched the pressure gauges. As the well filled, pressure rose and then dipped. Fractures were being created. Water was forcing its way into the shale. As the shale cracked and water rushed in, the pressure was dropping and then building up again. After an hour, he gave the order to gradually add sand to the mixture. He had waited to prevent yet another screen out.

It is flat in this part of North Texas. In 1929 Clyde Barrow—who later paired with Bonnie Parker—and his brother had robbed a bank in nearby Denton. When they couldn't open a small safe, they carried it outside and loaded it into the getaway car. They were spotted by police, and one of their accomplices was shot and arrested. Later, Bonnie and Clyde may have also tried to rob the bank in Ponder, two miles away from the S. H. Griffin #4. Local historians say they came away empty-handed. The bank had gone belly up a couple weeks earlier. But they didn't know what Steinsberger suspected. The real money around here wasn't in the banks. It was deep underground.

Over the next five hours, Steinsberger watched the water pressure. It would rise and then fall as the cracks elongated or the water broke into existing fractures, filling them up. Maybe that diploma would need to be eaten after all. But he still wasn't sure if this supersized frack job would work. Cracking the rock was only the first step. What would happen when the water was retrieved? Would gas flow out? And how much? These questions would be answered in the coming days.

In the early afternoon, he drove about forty minutes into

Fort Worth to Mitchell Energy's offices. His first stop was to see Mark Whitley, his supervisor and one of his allies within the company. Whitley's windows looked out onto a naval air station. The Mitchell engineers would watch the military test the jets' airworthiness. The pilots would fly the planes straight up into the air, go into a full stall, and drop before reengaging and darting off to the horizon. Steinsberger appeared in Whitley's doorway. The young engineer wasn't prone to bursts of emotion. Whitley had learned he needed to study his face to detect even the slightest glimmer of excitement. "Finally, we got one away," Steinsberger told his boss. The well had swallowed, in one long gulp, the mixture of water, sand, and chemicals. The slick-water frack had worked.

Over the next few days, Steinsberger and Whitley kept close tabs on the well. It played a cruel joke. The pressure kept dropping as more water came out. Whatever gas was in the well was weighed down by the eight-thousand-foot column of water. The water slowed to almost a trickle as it was pumped into nearby tanks. A couple days passed. Then the water began gurgling as bubbles of gas pushed up through the well. Finally, after about five days, almost all of the water was removed, and gas began to flow. The well was connected to a pipeline and measured. It was a monster well. Gas was screaming out.

A lot of the wells at the time that were fracked using gel would roar to life and then fall off very quickly, sometimes in a week or two. "This one didn't fall off. If a well made more than 70 million or 80 million cubic feet in the first ninety days, it was an 'A' well. This one made 1.3 million cubic feet a day for the first ninety days," said Steinsberger. "It was the best well we had ever had at that point. After this well, we knew we had turned the corner."

What Steinsberger accomplished that day was a dividing line for the energy industry and the country. Before the S. H. Griffin, engineers thought shale rocks were too dense to crack open. Gel could do the trick, but it tended to gum up the cracks, leaving no room for even tiny gas molecules to escape. Steinsberger—and most other fellow graduates of Texas and Oklahoma petroleum engineering departments—believed the United States was running out of gas. It had become a giant importer of oil. It was only a matter of time before it became a world-class importer of gas also.

The S. H. Griffin well began to change that thinking. Fifty-two years after Stanolind tested its first hydrafrac, Steinsberger had reinvented it and given birth to the modern frack industry. He had figured out how to force shale to give up its gas. There was now a bounty of energy sitting under American soil. Since this first well, more than a hundred thousand wells have been fracked in the United States. Every single one uses a technique similar to what Steinsberger first tried near Ponder, Texas. The era of the massive slick-water frack had begun. By today's standards, Steinsberger's well was small. He used 1.2 million gallons of water. Some modern wells use five times as much. There are other critical differences. Steinsberger's well went straight down. Most modern shale wells, like those in North Dakota, are "horizontal." They head straight down and then turn until they run parallel to the surface, traversing through the shale formation for up to two miles. But the breakthrough had been made. Steinsberger demonstrated that water could be used to create fractures in shale. Not only was it cheaper than using gels, it was better.

It didn't take long for word of the well to filter up to George Mitchell. He called Mark Whitley twice a week and asked the same two questions: "What is new?" and "Have you got your costs down yet?" Whitley told Mitchell about the S. H. Griffin, but tried to keep his boss's expectations under control. It was only the first well. Good sense dictated waiting for months to see how the well played out. But

Mitchell was not particularly patient. "If you tried not to tell George about something, because you thought he would go too far, it wouldn't go well for you," said Whitley. Mitchell liked what he heard and told him to keep going.

If a quiet excitement was percolating in Mitchell Energy's Fort Worth offices, people who lived near the S. H. Griffin had a different reaction. Robert Catron bought a piece of land on Tim Donald Road in early 1998. By the time he purchased a double-wide trailer and had it moved to his property, a ten-story drilling rig had been set up right across the two-lane road, about three hundred feet from his bedroom. It was drilling the S. H. Griffin well. "I was really disappointed. I had put my money down, and I didn't want to live across the street from this," he said years later in an interview. The truck traffic was overwhelming for a couple months. "It was a daily mess." He didn't even get the benefit of a nice royalty check. He owned the land, but a previous owner had kept the mineral rights. He had no idea the drilling rig was only the beginning. Several years later, he stood in his front door and counted eighteen wells being drilled. "This is nothing you want to have in your front yard," he said, "but there's nothing you can do about it."

On a drizzly morning in early 2012, I met Steinsberger in Ponder, outside the redbrick bank that, according to lore, Bonnie and Clyde tried to rob. It is now home to a high-end boot maker, but the old wooden cashier cages remain. Steinsberger's days of driving an old clunker are also over. He pulled up in a Lexus SUV. It was dark gray, the color of shale. I climbed into the passenger seat, and we set off for a brief drive over to the S. H. Griffin. As we neared the well, my map-reading skills failed me, and we made a wrong turn. Within a hundred feet, Steinsberger pulled into the entrance of a different gas well to turn around. He glanced at the sign identifying the well. He remembered it. "Some of these wells are more my kids than my own kids," he said.

After we started heading in the right direction, confusion set in again. When Steinsberger drilled the S. H. Griffin in 1998, there were no other wells nearby. But as we drove three-quarters of a mile on Tim Donald Road, wells kept popping up. The wells were in clumps, and I had a hard time figuring out how many were in each location. Steinsberger offered to count the transmitters atop each well. Each well had its own transmitter, and they were easy to see. As Steinsberger counted the meters, I tried to keep a running tally.

By the time we turned into the S. H. Griffin, we had lost count at around twenty-five wells. We parked at the entrance to the well pad. It is a two-acre flattened rectangle, with four olive green tanks used to store any liquids that come out of the well. There are two wells on the pad, both surrounded by chin-high chain-link fences.

"It's neat. I probably haven't been back here for ten to thirteen years," he said, as the car idled. "Before the S. H. Griffin number four, this was a completely uneconomic, very marginal area. To go from that to the hype we have now. So many wells have been drilled. Thousands. There are fifteen thousand wells drilled in that Barnett. That makes me feel proud."

The Barnett Shale was a riddle and a challenge. Steinsberger had approached it like a mathematician working on a nettlesome proof. There was gas in the shale. That was an article of faith at Mitchell Energy. George Mitchell himself instilled that belief. But the gas was trapped inside, and the rock was buried two miles underground. Complicating matters, Steinsberger couldn't even see what was going on in the shale. All he had were surface measurements. (Later, instruments that could measure and map the tiny fractures became commonplace. But in 1998, early attempts had failed. The heat inside the well fried the circuitry.) This sort of challenge makes petroleum engineers tick. Steinsberger's new well-completion technique was one of the most important technological breakthroughs of the twentieth century. Thousands and thousands of wells have been drilled into the shale. Ten trillion cubic feet of gas have been extracted. Soon after it was clear that the S. H. Griffin was a success, Mitchell started completing all its wells in the Barnett with slick-water fracks.

Steinsberger didn't set out to solve the problem of cracking shale. It found him. When he graduated from high school in Columbus, Nebraska, he attended the community college, where his father taught, for a year. He wanted to become a petroleum engineer, so he visited colleges in Texas before settling on the University of Texas at Austin. His timing couldn't have been worse. When he graduated in 1987, the industry was reeling from oil prices that were less than $20 a barrel. Jobs were hard to find, so he decided to backpack through Europe and Egypt for six months. While in Egypt, he was offered a job by the global oil-field service company Schlumberger. Being an overseas itinerant in the oil industry has its appeals, such as seeing the world and working in the most prolific oil fields. But it also means living in remote locations such as Egypt's western desert or a compound behind high walls in coastal Nigeria. Steinsberger wasn't interested. He returned to Texas and visited the university's career placement office. He put out resumes and got one bite—from Mitchell Energy. His starting salary was $30,000 a year. His first job was as a lease operator, a glorified babysitter, in Whittier, California. He was a thirty-minute drive from the THUMS Project that had set him on course to become a petroleum engineer.

Within a couple years, Steinsberger moved to Fort Worth. His bosses there recognized that he was a talented and inquisitive engineer. He sought out challenges, and by 1995, he was wrestling with Mitchell Energy's largest challenge: how to get more gas out of its

North Texas properties. The company had drilled about two hundred shale wells, and its executives had an appetite for another fifty. If there was no improvement, the plan was to spend the money on something else that showed more promise. Steinsberger decided the best he could do was save the company some money on these wells. Mitchell Energy and the rest of the industry hired oil-field service companies to perform the frack jobs. These companies charged by the gallon for the gel and threw in the horsepower needed to pump it down the well for free. Steinsberger began using less gel in each well to lower costs. The service companies charged a 1,000 percent markup on the gels, making them extraordinarily profitable. They warned Steinsberger that high temperatures at the bottom of the well would break down the lighter (and therefore less expensive) gels, rendering them unable to convey sand all the way to the fractures. What he found was the opposite. "The wells were just as good, if not better, plus I was saving thirty to fifty grand," he said.

Encouraged by these results, he began to wonder if he could do away with gels altogether. Water could be purchased from local cities for a fraction of the cost of gels. He wasn't sure if the water would work, but since it cost so much less to frack a well with water, he could improve the economics of the well if it was even close to producing as much gas as a well fracked with gel. What's more, if it didn't work, the water wouldn't clog up the shale. Mitchell Energy could always go back in with a gel frack.

The Fort Worth oil engineering community is tight knit and friendly. Several small companies had offices there. Engineers would often gather after work for beers and brisket at Angelo's, a barbeque restaurant on the west side of town where dozens of stuffed animal heads peered down on diners. They golfed together on weekends and gathered at regular symposia hosted by professional organizations to learn what was new. At these gatherings, Steinsberger learned that a Fort Worth company, Union Pacific Resources, had been experimenting with water fracks in East Texas sandstones.

In a 1997 engineering paper presented at a conference in San Antonio, Union Pacific's Ray Walker wrote about some of his successes. The conventional wisdom was that a viscous gel was required to transport the sand. If the gel was too watery, the sand would fall out and collect, uselessly, at the bottom of the pipe. The industry called these sands proppants. To show that gel wasn't needed and neither was so much sand, Walker and several coauthors titled the paper "Proppants? We Don't Need No Proppants." It was a cheeky allusion, he later explained, to the famous line "Badges? We ain't got no badges. We don't need no badges! I don't have to show you any stinkin' badges!" from the 1948 movie *The Treasure of the Sierra Madre*, starring Humphrey Bogart and Walter Huston. Walker and some colleagues described in the paper how they were cutting the cost of fracking wells in half, or more, and were still getting good results. There was just one issue. "Why it works is still generally unknown," Walker wrote.

Not that this mattered to Walker. Engineers are problem solvers. If the wells were cheaper and gas production better, problem solved. A later generation of geologists and engineers could worry about why. They were making better wells and improving their company's bottom line. As Walker tells it, he stumbled upon water fracks by accident. A gauge that measured water volume at a Union Pacific well had broken, and before the malfunction was discovered, a young employee had pumped in much more water (and relatively little sand) than planned. In a panic, the employee wanted to know what to do. "I said, 'Let's just flow it back and see what happens,' " he said. The well was a solid producer.

But Union Pacific was working with sandstones, not shales. There is a key difference between sandstone and shale. The ability of a fluid or gas to flow through rock is measured by geologists in dar-cies, named after the nineteenth-century French hydraulic engineer Henry Darcy. Imagine a large wave depositing water onto a beach. Some of the water will quickly disappear into the dry sand as it drops through channels. Beach sand has a measurement of about 5 darcies. The East Texas sandstones were 0.0001 darcy, or fifty thousand times less permeable than sand. The Barnett Shale is one thousand times less permeable than the sandstones. It is about 100 nanodarcies, or 0.0000001 darcy.

Steinsberger wanted to learn more about what Union Pacific Re-sources was doing, even though the company was drilling into such a different rock. He got permission to observe one of its frack jobs in East Texas. He then called up Walker and asked for a meeting. Walker thought that Steinsberger wanted to discuss a possible job at Union Pacific. Steinsberger had a different agenda. When he arrived at Walker's office, he took out a bunch of well data and maps of the Barnett Shale. Where did he think would be a good place to try a water frack? Walker was ecstatic. He even offered data and engineer-ing to help Steinsberger sell the idea to his bosses. It is hard to imag-ine Silicon Valley engineers helping a competitor pitch an idea for a new breakthrough design for a smart phone, but the Fort Worth community of petroleum engineers was collaborative. In their eyes, they weren't only trying to generate a profit for their companies. They had a higher calling. The United States was running out of en-ergy. Anything that could be done to reverse that trend was worth doing. A new approach to getting gas out of the rock would help the country *and* their companies.

After Steinsberger's four false starts, the S. H. Griffin first showed that using a massive slick-water frack would open up shale. Still, the idea was so revolutionary that it took time for the rest of the indus-try to accept it. Many engineers simply weren't ready to concede that it was possible to frack shale rocks and get gas out. "I ignored it," recalled Kenneth Nolte, an engineer at another oil company who spent his career on smaller fracks targeting sandstones and other more porous rocks. He couldn't wrap his head around fracking shale. Shale was an impermeable barrier, not a source of gas. "It is just as startling as saying ice doesn't freeze anymore," he said.

At first Mitchell Energy decided to keep quiet and not broadcast its success. But internally, the company welcomed Steinsberger's breakthrough. By the end of 1998, six months after the S. H. Griffin, it was using slick-water fracks for every well it drilled. And it began going back into old wells fracked with gel and refracking them with water. This technique also worked. Under oil industry accounting, the value of an exploration company correlates closely with its proven reserves. These are oil and gas deposits that have been discovered and can be extracted with existing technology and current prices. If the company could go into hundreds of already drilled wells, add water, and raise its production, under accounting rules it could add around 750 million cubic feet of gas reserves for all of its wells. Steinsberger had literally increased the value of Mitchell Energy by a couple billion dollars.

Steinsberger told me he didn't receive a raise or any bonus. George Mitchell never called to congratulate him. But he no longer worried about getting fired. Before the S. H. Griffin #4, the industry had been fracking its wells for nearly five decades. It had used hydrochloric acid, nitroglycerin, napalm, thick gels, and even nuclear bombs. Steinsberger showed that there was a simpler way: water, lots and lots of water.

A few years later, Steinsberger set out on his own, a fracker for hire. He estimates he has had a hand in drilling more than one thousand shale wells. Several small operators have hired him to work on the shale developments in Texas, Pennsylvania, West Virginia,

Alabama, and Arkansas. He'll go anywhere in the United States to drill wells, he said, except for the booming North Dakota oil field. "Too cold."

After we left the S. H. Griffin, we drove around the area. Steinsberger talked a bit about plans to frack some wells in the famed Permian Basin. An enormous geological area that covers much of West Texas, it contains some of the largest oil fields ever discovered. Until a few years ago, it was considered played out. The oil had been extracted. Big oil companies had sold off the leases, and production had plummeted. But in recent years, companies have returned to the Permian Basin. Using fracking, production is rising again. Fracking is no longer just to get at gas. In the Permian, the target is oil—good old-fashioned Texas tea.

We headed south on Tim Donald Road, past the twenty-five or so wells. Another half mile on, we passed a large industrial facility that looks out of place amid the houses and ranches. It belongs to Chesapeake Energy. There is a two-story beige building that strips liquids such as ethane and butane—used by petrochemical plants to make plastics—out of the gas. There are a couple large compressors that send the gas into a pipeline for delivery to the Gulf Coast. A minute later, Tim Donald Road ends, taking a tight turn to the right. Ahead of us was a new drill pad behind a private house, the earth flattened, packed down, and ready for the drilling rig. In the distance, a rig was at work, lights blinking up and down its steel scaffold to alert any low-flying airplanes to its presence.

On the morning we met, it had been nearly fourteen years since Steinsberger fracked the S. H. Griffin. Gas was still flowing from the well. It had produced 2.3 billion cubic feet of natural gas. While prices for gas have risen and fallen over the years, it has generated about $11 million worth of gas. There's likely another half billion cubic feet still to come, although each year the amount declines.

Most of the gas that a shale well will produce in its lifetime comes out in the first year or two. After a couple years of dramatic declines, the well will slow down and decline for years. How long shale wells will produce is a matter of some controversy in the energy industry. If a shale's tail, as it's called, is long and its decline slow, the wells being drilled will generate gas for decades to come, creating a nice flow of fuel for the country. But Wall Street wants energy companies to increase production year after year. And since their shale wells might decline by 60 percent in the first year and another 30 percent in the next year, these companies need to keep drilling hundreds or thousands of wells just to keep production flat.

This demand helps explain why there are so many wells in and around the original S. H. Griffin. This drilling has created a glut. There isn't enough demand for the fuel. Yet wells are still being drilled here in the Barnett, even though, with natural gas fetching less than $4 for every thousand cubic feet, it is not clear if these wells will make enough money to pay for the cost of drilling them.

"Our country has to develop more ways to use natural gas. That is just the way it is. We are not going to run out of natural gas in our lifetime or our kids' lifetimes for sure," Steinsberger said. "Are we going to export natural gas? Or are we going to try to find more uses so that we can try to diminish use of coal and oil?" He admitted that he doesn't spend too much time thinking about such policy concerns. He's an engineer and focuses on the job at hand. Where to drill? How can he design the optimal frack job? How can he keep costs low?

We made our way back to Ponder and parked at a gas station cafe with a couple Formica tables and chairs. We grabbed coffees and talked over country music and conversations at nearby tables. He still marvels that the industry found so much gas from reservoirs that were overlooked just a few years ago.

"A lot of people still think you drill a well into a big lake of gas or oil, and that is how it comes out. That's not true," he said. "The Barnett is the tightest thing in this building except for the metal. It is harder than that cement. It is harder than anything else in here. It is a very tight formation. It is hard to believe oil and gas can come out of that."

7

LARRY WAS THE BRAKE

Nick Steinsberger's breakthrough in 1998 couldn't have come at a better time for Mitchell Energy. The Mitchell family was facing both personal and professional crises. The price of oil was collapsing, dragging down the stock price of Mitchell Energy. Mitchell owned a majority of the publicly traded shares of Mitchell Energy, and most of his wealth was wrapped up in the stock. He had pledged sizable donations to the Houston Symphony, a performing arts center in the Woodlands, and a think tank he had created by borrowing from a syndicate of ten banks. His Mitchell Energy stock collateralized the loans. As the stock price fell from $35 to $10, the banks clamored for more collateral. Todd Mitchell, his son, tried to keep the bank syndicate from foreclosing the loan and seizing Mitchell Energy stock. If the banks had demanded the stocks, George Mitchell would have lost control of his company. At the same time, George

Mitchell was being treated for prostate cancer, and his wife, Cynthia, was showing early signs of Alzheimer's disease.

As news of the new fracking success filtered up to Mitchell and the board, they set a plan in motion. The company would double its pace of drilling Barnett Shale wells, increase its value, and then find a buyer. By 1999, the Mitchell family had decided it was time to get out of the oil and gas business. It hired Goldman Sachs to sell the family firm. No one wanted to buy it. The company had achieved a technological breakthrough that was so revolutionary and violated so many basic tenets of finding oil and gas that workaday engineers weren't prepared to believe it. A data room was set up in Houston where prospective buyers could look at a wealth of information on Mitchell's wells and acreage. A few companies visited it, but left befuddled.

Mitchell's timing was terrible. Oil prices were low, and big oil companies were focused on megamergers to drive down costs and put them in a position to compete for giant, unimaginably complex overseas projects that often cost tens of billions of dollars. BP kicked off this consolidation spree in 1998 when it bought Amoco and then, a few months later, Arco. Exxon swallowed Mobil in 1999, reuniting two of the largest pieces of Rockefeller's Standard Oil. A year later, Chevron and Texaco joined together, and Conoco and Phillips Petroleum joined a year after that. Wall Street rewarded these marriages. With the exception of big deepwater fields in the Gulf of Mexico, analysts and industry engineers considered the United States picked over and used up. A Big Oil executive proposing buying a gas field in North Texas not only would have been swimming against the current but also would have been putting his career in jeopardy.

Even the smaller energy companies that focused on the United States weren't particularly interested. Jeff Hall, part of a Devon

Energy technical team that assessed potential acquisitions, traveled to Houston to examine Mitchell Energy. He was underwhelmed. "My view was, 'Eh, this is kind of a marginal deal,' " he recalled. "It looked like a bunch of old, tired conventional production." Even the new Barnett Shale wells didn't pique his interest. He remembers thinking these wells took a lot of work and a lot of money. And for what? He returned to Oklahoma City and prepared to brief senior management. Devon was a company that liked to purchase other companies. Hall's job was to be the naysayer and dowse the company's acquisitive enthusiasm with cold water when the data didn't warrant it. He came to the meeting to discuss a Mitchell acquisition with a bucketful of ice-cold water. One of the stumbling blocks, Hall told Devon's chief executive, Larry Nichols, and other managers was that no one on the assessment team understood how the gas was being produced out of the shale. "We turned up our noses because we didn't think it would work," Nichols recalled. In April 2000, after a few months during which the data room drew as much interest as a clunker on a used-car lot, Mitchell Energy ended the sale.

Mitchell decided to double down and prove to the industry that his new shale wells were worth a second look. It began reopening its old wells and fracking them with its newfangled slick-water approach. If it worked on new wells, maybe it could perk up older wells too. The results were good. Gas production in some of the reopened wells jumped tenfold. What's more, it placed tiny microseismic sensors in the ground to provide evidence that it was creating new, longer fractures. At a conference of petroleum engineers in Dallas held in October 2000, Mitchell boasted that reopening old wells could "penetrate untapped sections of the reservoir, significantly increasing production rate and reserves." These refracks cost more money, he tried to explain, but paid for themselves with more gas.

In May 2001 Mitchell Energy asked its bankers to approach

Devon again. Show the company the new wells, Mitchell suggested, and see if it can be tempted. At about the same time, Nichols attended an energy investor conference where he wandered around the hotel foyer looking at other companies' promotional materials. He picked up a Mitchell Energy presentation and was struck by rising gas production. Meeting in his office, Nichols asked Hall and other members of the team that had evaluated Mitchell what was going on. They repeated the same message from a year before: fracking shale wouldn't work. Nichols pointed to Mitchell's gas production and asked, "Why is it going up? Simple question." Nichols sent Hall back to Houston again, despite some grumbling about spinning wheels. By this time, the earliest Barnett Shale wells had a couple years of production data and looked better than he expected. Moreover, the refracks of old wells looked quite promising. Mitchell wasn't trying to put lipstick on a pig, Hall began to realize, he was selling a prizewinning hog. Hall, a geologist who had worked for wildcatter-turned-corporate raider T. Boone Pickens before ending up at Devon, got increasingly excited the longer he crunched numbers in the data room. He returned to Oklahoma and started working out some rough estimates. If these wells could be replicated across all of Mitchell's acreage, what would that mean? "This was just a big gas factory," he concluded.

Before he presented his new thinking to Nichols, Hall decided to learn all he could about these shale rocks. There wasn't much in the petroleum libraries. "It could fit in my briefcase. There were three books and a few publications," he said. This was a little nerve-wracking, but production numbers don't lie. Mitchell Energy was getting a lot of gas out of these rocks. When the time came to make a recommendation to Nichols and other executives, Hall's message was different from a year and a half earlier. "This looks like a real good opportunity," he told his bosses. Expecting Hall to advise against a deal, the management team sat in silence for a few seconds. Then Nichols spoke: "There's a deal to be done here. Let's go after it."

Devon had bought many companies, and Nichols knew how to close a deal. There were teams of employees ready to conduct due diligence. Devon moved quickly, worried that someone else might see what it now saw in Mitchell. Within three months after the bankers approached Devon for a second time, the outlines of a deal came together. But another realization tempered Nichols's enthusiasm. Mitchell had been drilling vertical wells. Extrapolating the amount of natural gas available from these wells made the $3 billion cost to acquire Mitchell a "full price," said Nichols. However, Devon had used horizontal wells elsewhere and thought that these wells, which descended vertically and then bent underground to run parallel to the surface, might be able to access more gas. But it was all paper calculations. Mitchell had tried a handful of horizontal wells without much luck. There was no way to know if horizontals could turn a Mitchell acquisition into a great deal until Devon could try a few. It would have to wait until it owned Mitchell's assets.

Devon made an offer and then bumped up the price it was willing to pay after Mitchell Energy's bankers said the initial offer wasn't enough. The bankers again asked for more, and Nichols balked. Its bluff called, Mitchell Energy acquiesced. Lawyers and bankers worked through the weekend at the law firm Vinson & Elkins's offices in a Houston skyscraper, finalizing details. On Tuesday, August 14, 2001, before the stock markets opened, both companies issued press releases announcing the deal. Devon's stock rose that day as analysts absorbed details of the transaction and concluded it had bought a lot of gas on the cheap. A month later, after the terrorist attacks on the United States, Devon's stock would fall along with the price of oil and natural gas, making many in its Oklahoma City headquarters nervous. But it was a temporary blip. Better days for the company were ahead.

Larry Nichols was born to be an Oklahoma City oilman but tried his best to avoid his fate. When he was ten years old, a prolonged drought prompted his father to drill a water well in the backyard of the family home. The future CEO was fascinated by the drilling rig. He spent the afternoon watching the work. When it started producing water, he ran into the house and exclaimed, "They've struck oil, but it looks like water to me!"

The Nicholses were a well-to-do, respected clan, part of the Sooner State elite. His father, John W. Nichols, was an accountant and a financial genius with an inclination to use every conceivable trick in the book to avoid paying taxes. When there weren't enough tricks, he invented new ones. After graduating from the University of Oklahoma, he went to work on the New York Stock Exchange in the midst of the Great Depression. He returned to Oklahoma City in 1936, humbled but with a firm understanding of tax law.

In November 1950 he created the world's first drilling fund registered with the Securities and Exchange Commission. Before this time, money to drill wells was often raised from wealthy investors on handshake deals. The oil business was rife with swindles and outlandish promises of easy money. The SEC regarded the formalized drilling fund as a way to protect small investors. But John Nichols wanted to tap the wallets of some of the richest men in America, not small investors.

Less than two months after the SEC approved the new fund, the thirty-five-year-old Nichols convened a meeting at the University Club of Chicago. In a private room, several wealthy industrialists gathered with their accountants and lawyers around linen-draped tables. The president of the Joseph Schlitz Brewing Company was there, as was the chairman of Bethlehem Steel, and meatpacking magnate Phillip Armour. You are "plagued" by the 90 percent federal tax on income, Nichols told the assembled members of the corporate elite. His new drilling fund exploited the tax code's allowance for a well's expenses to be deducted in the year it was drilled. Expenses often made up 70 percent of the cost of a successful well, he explained, allowing for investments to generate large tax deductions as well as ongoing income from the wells. The result was little taxable income for the investors. "Nichols explained how his plan could almost miraculously save high-income investors up to half their annual income tax bill," wrote John Nichols's biographer Bob Burke in *Deals, Deals, and More Deals*. Years later, Congress would close the loophole that Nichols had opened, but at the time, it was all perfectly legal and rather ingenious.

In broad terms, here's how it worked for a fund that raised $1 million. About 40 cents of every dollar raised through Nichols's investment fund was used to drill wells in counties where there was already significant energy production. Three in four wells would strike oil and gas. The cost of drilling the three successful wells would be about $300,000 and generate tax credits of $240,000. The $100,000 spent on drilling one duster was lost. The remaining $600,000 was used to drill wildcat wells. If those six wildcats were dry, it created another tax credit for writing off the value of exploration. For rich investors with a lot of income, they could generate plenty of tax credits to lower their IRS bill and gain a potential dividend from future production out of the successful wells. Investors in Nichols's first fund included the Badenhausen family, which owned the P. Ballantine & Sons brewery, as well as top executives from both Chrysler and General Motors. The Pillsbury family also invested, as did Hollywood stars Ginger Rogers and Barbara Stanwyck. The first fund delivered the tax credits as promised, and the technique began to spread. By the 1980s, when a change in tax law ended these drilling funds, billions of dollars had been raised through the investment vehicle pioneered by Nichols.

After raising money from the nation's entertainment and industrial elite, Nichols signed up smaller but still quite affluent investors with promotional sales brochures. On the cover of one, a picture of an oil well blows a black cloud of oil into the sky. "Oil: The Last Frontier for High Tax Bracket Wealth," the brochure crowed. Inside was more heady prose: "The arithmetic of riskless investment" and "Big projects for big profits." His son Larry would later chuckle about his father's sales strategy, telling a group of Oklahoma businessmen in 1996, "The SEC would send you straight to jail for that now."

Larry Nichols graduated from Casady School, a relatively new day school formed by the Episcopal Diocese that became the place for Oklahoma City's elite to educate their children. From there, he studied geology at Princeton University. Then he decided to leave the oil track and headed to Ann Arbor to complete a law degree at the University of Michigan. A top student, he served as an editor of the law review and clerked for Supreme Court Justice Tom Clark. Not long after Nichols arrived in Washington, DC, Justice Clark resigned from the court after his son, Ramsey, was nominated to be US attorney general. Nichols transferred to the office of Chief Justice Earl Warren for the remainder of his clerkship. This new assignment brought little meaningful work, and he drafted no opinions. When his clerkship ended, he joined the US Justice Department as special assistant to Assistant Attorney General William Rehnquist. In 1970 his father lured him back to Oklahoma City with the promise of creating an oil company together. Larry Nichols decided to try it for two years. He figured he could always return to Washington, DC, and a profitable law career. He got out of the capital just in time. A couple years after he left, the Watergate break-in and cover-up began, a scandal that would reach into the Justice Department offices where he had worked.

The company that John Nichols built was called Devon International. Like his earlier work, it was craftily constructed to avoid taxes. The parent company was incorporated in Luxembourg, which had no income tax on corporate profits. Nichols set up a subsidiary in Oklahoma to own oil and gas properties in the United States, and profits flowed through Saxon Oil, a Panamanian corporation, and a United Kingdom subsidiary. In a nutshell, the complex structure allowed rich Europeans to invest in US oil and gas properties at a low tax rate. "Financial creativity," Larry Nichols said later, "is part of our heritage." Larry served as the new company's in-house lawyer.

If John Nichols was an innovator, ready and willing to take risks to create wealth and tax shelters, Larry Nichols tended to be careful. With a lawyer's conservative bent, he focused on making sure that the new Devon didn't succumb to the wild vicissitudes of oil and gas prices. Mary Nichols, John's wife and Larry's mother, summed up the longtime working relationship that emerged between the two men. "John was always the dreamer. He could figure out new deals, and Larry, often taking a devil's advocate position, could tell him if they would work. John was the accelerator. Larry was the brake," she said.

In 1986 Ronald Reagan signed into law a simplified federal income tax. One change ended the investor-backed drilling funds that John Nichols had pioneered. Shutting this loophole could have been catastrophic for Devon, but Larry Nichols believed that it created a new opportunity. The end of the tax-advantaged drilling funds dried up a large source of funding for many of the four hundred publicly traded oil and gas companies in the United States, many of them small. Larry Nichols, already the CEO of Devon, realized that many of these companies wouldn't be able to survive. The oil industry would enter a period of consolidation. He wanted to be a consolidator, picking up assets from distressed companies and swallowing other troubled companies entirely.

Within a decade, Devon had doubled in size and then doubled again. It developed a reputation as a ravenous acquirer of assets. Larry Nichols became known as an oilman who spoke softly and carried a large checkbook. Devon also developed a reputation as a company that could handle unconventional assets. Devon acquired a large acreage position in the San Juan Basin in northern New Mexico. The area had been drilled for decades. Part of John Nichols's first drilling fund, in 1951, had been used to acquire leases and drill in the basin. These older wells had targeted a deep rock formation. Above that was the Fruitland coal formation. There was a lot of gas in the coal seams, but also water. Three companies hit upon a new approach to this so-called coal-bed methane: Amoco, Burlington Resources, and Devon. They would create a cavity in the coal and pump out thousands of gallons of water. After the water pressure was lowered, gas would flow out of the coal. It was laborious work, but it was supported by a nice tax credit that the government had created to help companies develop new gas sources.

Of the three companies, Devon was most intent on acquiring new acreage in the San Juan Basin. When property came up for sale, by the time Amoco's team in New Mexico had sent the particulars of the deal to headquarters in Chicago for consideration, Devon had already closed the deal. "That we could compete with Amoco was a revolution in my mind," Larry Nichols said. "We could move faster on little deals, right under the nose of the big guys." The lesson that he took from the San Juan Basin was that Devon could compete with anyone, even a progeny of Standard Oil. He also realized that oil-field technology was being democratized. In the past, Big Oil companies such as Amoco had a lock on research and development. They patented and controlled the major breakthroughs. But most of these companies had slashed their research budgets. The technology game was much more wide open by the early 1990s. Even a middling Oklahoma City company could compete on cutting-edge exploration that required a new approach and new technology. He would apply this lesson in the Barnett Shale.

Before Devon even closed the deal for Mitchell, it sent some of its engineers to embed with the shale operations and get a better idea of what was going on. Brad Foster, head of midcontinent operations for Devon, was given control of the Barnett Shale. He grew up in Pittsburgh and, in the late 1970s, during summers while in college in West Virginia, he worked for a local gas company that tried a couple fracks into shales. He has two recollections from those summers. The first: "How does anyone make any money doing this?" The second: "I will never work a shale project again." Just the guy to be handed the Barnett Shale.

Foster faced a tricky situation. Mitchell Energy had been drilling a lot of plain vertical wells, straight down, and then using its new fracking technique. This practice worked well enough in a portion of the Barnett Shale where the rock sits above limestone. But there was a much larger area where there was gassy Barnett, but no limestone. Underneath the shale was the Ellenberger formation, rocks riddled with salty water. When wells in this area were fracked, the slick water injected into the well came in contact with the salty reservoir. Most of the energy of the frack—the massive horsepower assembled to force in millions of gallons of water—would dissipate in the Ellenberger. Instead of creating large man-made fractures to drain the

Barnett, it would create small fractures into the Ellenberger. The result was expensive salt-water wells. If Devon wanted to create a replica of the Dead Sea in northern Texas, these wells were ideal. For any other purpose, they were duds.

Devon had a different approach in mind, one that Mitchell had gingerly attempted years before with federal research grants. The idea was to drill horizontal wells in the Barnett. A vertical well is like an elevator that picks up passengers at every floor. The Barnett is typically 350 feet thick, or the size of a thirty-five-story building. Devon's horizontal wells would traverse through the shale, running a couple thousand feet. This elevator, laid on its side, would be longer than the world's largest skyscraper. Devon's horizontal wells would reach much more of the Barnett gas from a single well. There was another potential major advantage to horizontal wells. Devon hoped that these wells would help it avoid the Ellenberger's salty trap. If the horizontals didn't work, Nichols knew he had paid a hefty price for Mitchell. But if Devon could frack where the Barnett sat above the Ellenberger, the number of wells it could drill on Mitchell's acreage would triple or quadruple. Buying Mitchell would turn out to be a steal.

One day in 2002, Nichols and Foster sat down to talk about plans for the Barnett. "I want you to expeditiously but carefully see if these horizontals work," Nichols told Foster. He warned, "Don't bet the family farm." Once a month, Foster's new Barnett team gathered in an unadorned conference room on the seventh floor of a downtown boxy building with a gray, bland lobby that looked more like it housed a collection of small accounting firms than the global headquarters of a company listed on the New York Stock Exchange. The only indication that the building housed an oil company was a model in the lobby of an offshore drilling platform.

Members of the Barnett team brought maps and printed-out

spreadsheets that tracked costs and gas production trends, as well as the time to drill and complete wells. Foster has thick fingers and a balding pate that make him look like Tony Soprano, but he has a warm personality. He cracked jokes and emphasized the need to experiment, but to try new approaches carefully. It "was about as far from full steam as you could get," said one attendee. Soon Foster turned the talk to horizontal wells as a way to get more out of the Barnett. "Why can't we bring this technology here?" he asked. Many of the field engineers, the people who actually drilled the wells in the Barnett, were holdovers from Mitchell Energy and resisted the idea. They had tried these wells and didn't think they would work. Some worried that an expensive failure would blot their careers. "Here's the deal," Foster told the team. Referring to Devon's management, he went on, "There are no repercussions. We are going to take all the accountability on this, and we're going to take the responsibility."

So, Foster continued, what are the obstacles? An engineer responded that there were no rigs in the Barnett that could handle these kinds of wells. A few days later, Foster and a couple members of his team drove to Tulsa to meet with Helmerich & Payne, a rig outfit. He told the company that he wanted a rig capable of horizontal drilling in the Barnett. "They looked at us with their eyebrows raised," he said. "What are you up to?" an H&P executive asked. Foster explained his theory. If the well could be turned to run directly through the Barnett Shale, perhaps the fractures wouldn't escape into the Ellenberger and produce salt-water wells. "We think it will work," Foster said. H&P agreed to send one of its most modern rigs to Fort Worth on a trial basis.

Horizontal wells, in 2002, were fairly unusual. Only one of every fourteen wells drilled in the United States and Canada was horizontal. (A decade later, six of every ten wells were horizontal.) While these twisting wells were still relatively new, the industry had been drilling slant wells for decades. In 1941 oil was discovered under the Oklahoma Capitol building. Even in oil-crazed Oklahoma, tearing down the steel-reinforced concrete dome was out of the question. Instead, Phillips Petroleum set up a drilling rig across the street. The well proceeded at an angle to travel underneath the building, and for years lawmakers met atop an active oil well. The well, nicknamed the Petunia #1, was drilled from a flower bed.

In 1986 Al Yost, the government scientist in West Virginia, and colleagues drilled a horizontal well into the Devonian Shale in southwestern West Virginia's Cabwaylingo State Forest. This well descended for 2,113 feet vertically, and then, using pipes bent at a slight 2.5 degree angle, the drill bit proceeded slowly to the right. The pipe got stuck once, and the angle achieved was too small, forcing them to retreat a few feet and start a new shaft. The motor that drove the drill plugged up several times and broke down every ten hours, on average. But over the distance of nearly six football fields, they reoriented the well until it ran horizontally, parallel to the ground for 2,000 feet. Viewed at a remove, the well resembled a truncated capital J. The well dipped into and ran through the shale, exposing more of the rock to a rudimentary frack. Yost reported in a paper that after fracking it, the well flowed at a rate seven times higher than vertical wells in the area.

Since that West Virginia well, the industry had gotten better at horizontal drilling. It can be done faster and with more precision. Devon drilled its first horizontal Barnett well in June and July 2002. The Veale Ranch #1H took nearly a month and a half to drill and was fracked with 1.2 million gallons of water. The well worked, although production was only marginally better than a vertical well. Five months later, in November 2002, Devon had its sea legs under it and was gaining both confidence and speed. Devon drilled its sixth horizontal well, the Graham Shoop #6, in less than a month. It used more than twice as much water as the Veale well to frack it, and more than seven times as much gas came out. In Foster's eyes, it was a beautiful well. His misgivings about drilling shale began to melt away. But when he asked his team at the monthly Barnett meetings to explain to him why the wells were so good, blank stares met him.

Shale wells are overachievers. The new fractures free up a lot of natural gas, which will rush into the well and up to the surface. The first few weeks that the wells are connected to a pipeline yield the highest production, and then they start to decline. A well will continue producing gas long after the gas freed up by the initial fractures travels up into the well. This gas appears to come from the shale rock itself, worming its way out and into the fractures. How this works remains not entirely understood.

Hundreds of engineers filled out an informal survey at a shale gas conference in November 2008. "I am confident that I understand reservoir drainage" was one question. Four-fifths of the engineers responded that they weren't confident. Several years into the all-out juggernaut of shale drilling, the experts themselves didn't know exactly why shale production worked. They just knew it worked.

If these engineers were puzzled about what was going on in the shales, what hope did Brad Foster have? In early 2003, he started to get production reports from Devon's first horizontal Barnett Shale wells. They looked promising. But he wasn't ready to pop any champagne. "We were a little bit on the pessimistic side," he said. "We were not sure we totally understand this shit." The world's first modern frack well, the S. H. Griffin #4, was all of five years old at that point. It had started off producing more than one million cubic feet a day but had dropped off to a quarter of that by mid-2003. Who knew where it was going?

Larry Nichols said later that the company was acting with what he characterized as "excessive conservatism." The lack of data led Devon to inch forward, instead of break into a run. "What we wanted to see was, Okay, it peaks, and we know it goes down at a very steep rate; where does it flatten out? If it flattens out at one level, you make a lot of money on that tail. Or does it continue on down and flatten out at a lower level?" he explained. Until time passed and he had real production data, he decided to "tread water" for a couple years.

Devon was following a time-tested method: drill a little, wait a little, drill a little, wait a little. In technical terms, the company was asking whether the decline curve would be exponential or hyperbolic. Exponential decline means that production falls off until it just peters out. Drawn on a graph, it looks like a straight line. Hyperbolic decline features a dramatic drop followed by a flattening out. If the declines were hyperbolic, as Foster thought they would be, then shale gas could be profitable under the right circumstances. But until he knew for sure, neither he nor Larry Nichols wanted to risk too much money drilling the Barnett.

In June 2002 Devon held a Barnett coming-out party for Wall Street analysts in a Dallas suburb. There were 1,043 wells in the Barnett, it explained, but there was room for another five thousand. It was using fourteen rigs in the Barnett. At that pace, it had a fifteen-year backlog of wells to be drilled. Not only wasn't Devon in any hurry, it had planned exploration around the world that would take a lot of money to finance. Ramping up spending in the Barnett wasn't a top priority. There was no need, Devon thought, to go out and find other shales similar to the Barnett. There weren't any. The company declared that it was "unique." It couldn't have been more wrong.

That fall, I decided to take a drive around the Barnett to see for myself what I had been hearing about in corporate presentations. Back then, you had to go looking for signs of gas development. The rigs were few and far between. I drove to a slight rise so that I could gain a vantage point to see above the strip malls and chain restaurants and find a rig. I saw one off in the distance and spent the next half hour taking wrong turns down cul-de-sacs until I found it. It had been erected in a field a couple hundred feet behind a cluster of suburban houses. Neighbors said it was loud. That it was lit up like an oversized Christmas tree at night annoyed them. This type of close proximity between homes and drilling pads was becoming the new normal. For the first time since the early 1900s, when oil extraction in Los Angeles began, an urban drilling campaign was under way.

In July 2003 Devon offered me a tour to show off the Barnett. I met a couple company employees at a Shell station off Interstate 35. Nearby, several trucks idled, waiting their turn to fill up from a metered municipal fire hydrant. We visited a drill pad and a large plant that took the gas and stripped out ethane, butane, and propane. Fracking hadn't entered the national discussion. The tour was a way for Devon to talk about its horizontal wells and demonstrate how it was getting gas out of the ground. Fracking wasn't a curse word. Not yet, anyway. Devon didn't even call it shale gas back then. My tour guide, a former Mitchell employee named Jay Ewing, called it "unconventional gas" to separate it from the normal way of drilling for the fuel. Years later, I met him again and spent another few hours driving around the Barnett. On the second tour, he showed off the three-story-tall sound walls that Devon built around its drilling pads to cut down on the noise and be a better neighbor. This time he told me his full name. My first official tour guide to the new world of fracking was none other than J. R. Ewing.

Back in the early days of the Barnett, Devon was far and away the largest holder of gas exploration leases in and around Fort Worth. It wasn't, however, the only one. There were several small companies making a modest profit on the margins of the old Boonsville Bend gas field. It was soon clear that fortune had smiled on them. They were in the right place at the right time. They held thousands of acres of drilling rights, obtained when leases could be had for a few dollars per acre and held in perpetuity by dint of an existing well that dribbled out a couple hundred cubic feet of gas a month. These lucky few became the first multimillionaires of the shale boom.

Dick Lowe, owner of a small local explorer called Four Sevens Oil, told me his business strategy was nothing more complex than to copy Devon. Under Texas rules, Devon had to file monthly statements about the wells it drilled and how much gas they produced. When Lowe saw Devon's production figures, he followed its lead. "As soon as we saw what their horizontal wells were doing, we started drilling all of our wells horizontally," he said. This was in 2003. Three years later, he sold the company for $845 million.

Chief Oil & Gas also rode on Devon's Barnett coattails. One day in 2005, company owner Trevor Rees-Jones explained the secret of his success. He was a small wildcatter who stayed close to home. Until the Barnett came along, he drilled and operated a lot of wells near Fort Worth. On a tour of his Dallas office, he explained what happened by pointing to a wall near a copy machine. "I was banging my head against the wall," he said, "and one day the wall gave in." He soon sold the company's Barnett holdings in a series of deals worth nearly $4 billion.

Before Devon, these companies had copied Mitchell and leased aggressively, creeping closer and closer to Fort Worth until they headed into the city itself. Sarah Fullenwider, the city attorney, began to grapple with how to zone these wells only a few months before Devon bought Mitchell. She had moved to Texas a couple years before from North Carolina, where oil and gas regulation wasn't on the state bar exam. She called around the country searching for other cities with ordinances she could copy. Meetings on the rules continued into the fall, including on the afternoon of September 11, 2001. Everyone was in shock and thought that carrying on normal work might help keep city workers calm, Fullenwider recalled. Finally, in December, officials passed a local ordinance requiring, among other things, that companies build an eight-foot-high masonry wall around a well within six hundred feet of a school, house, or park. The new rules didn't even merit a front-page article in the local newspaper. Fullenwider was surprised by how congenial the process was. "People were just used to it," she said. "It's just Texas." Local drillers were happy with new rules. Lowe told me that nothing was off-limits. "We could drill a well in our parking lot. We could drill a well on the courthouse steps. We could drill a well in the middle of TCU [Texas Christian University] Stadium," he said.

Neither Four Sevens nor Chief held patents that made them valuable. Neither company had any proprietary technology process, made a better widget than competitors, or had smarter engineers. They were bought for one thing and one thing only: their acreage. Four Sevens held 39,000 acres in the Barnett, locked up by existing production. New Barnett wells could be drilled on this property without difficult negotiations with the landowners. Chief held 169,000 acres. These were real estate transactions, not energy deals. The domestic energy industry was turning a corner. In the past, success came from an ability to find the biggest buried troves of oil and gas. Geologists used superstition (drill atop cemetery hills), hard work (George Mitchell poring over the squiggly lines of well logs late into the night), and, later, supercomputers processing seismic data to use sound waves bouncing off rock to find "bright spots" that indicated oil deposits. But in the Barnett, the gas was everywhere.

Some parts of the shale were better than others, held more gas, or were thicker, but wells found gas. In this new world, companies thought the key to success was speed. How quickly can you lease up thousands of acres atop a shale? This was the job of men (and a few women) rifling through files in county courthouses to research who owned the gas rights, and then knocking on doors to get leases signed. A military nomenclature emerged. Companies deployed armies of these "landmen" to capture acreage. A land war began.

Devon's genetic makeup didn't fit well in this new energy world order. It was too cautious. It could move fast to make a deal, but then it would become conservative when it came time to lease and drill. It preferred to spend time driving down costs. But if Devon was content to take it slow, a crosstown competitor would soon begin taking a different approach. Before Chesapeake Energy began investing billions of dollars to snap up every drillable acre it could find and kicked the shale boom into overdrive, the company needed to see this newfangled gas production firsthand. Its introduction was accidental.

A week after Devon's 2002 presentation to Wall Street analysts, Chesapeake acquired Canaan Energy. Chesapeake CEO Aubrey McClendon had been pursuing Canaan for more than a year. When Canaan rebuffed McClendon's initial offers, he used bare-knuckled tactics pioneered by corporate raiders in the 1980s. "We believe it is clear that management's plan is not working," McClendon wrote in an open letter to Wall Street after buying up 7.7 percent of the company's stock. "If given the opportunity, most Canaan shareholders would prefer to sell their stock at a premium to us rather than waiting on management's plan to work." Canaan capitulated.

McClendon wanted Canaan because of its wells in western Oklahoma. The two companies' wells were so close together that Chesapeake workers heading out to service their own wells would drive past Canaan wells. Bringing together the companies could reduce costs. Now that he owned Canaan, the brash executive needed to figure out what to do with another small Canaan asset, a minority stake in some exploratory acreage south of Fort Worth. In the Canaan deal, the Barnett position was "an afterthought," McClendon said, and "probably a liability." He decided not to do anything with it. Hallwood Group, a Delaware holding company with energy assets, owned the majority stake in the partnership, which meant that it got to call the shots and decide where and when to drill. Chesapeake went along for the ride, writing checks for its share of costs when Hallwood sent invoices. Hallwood was an ambitious and technically competent operator. It copied Devon's horizontal wells and began increasing the size of fracks. It also ventured into Johnson County, slightly south of Fort Worth, where there were no rocks to keep the fracks from breaking out into the Ellenberger and creating giant saltwater wells.

Bill Marble, a Hallwood energy executive, described the challenge in heroic and somewhat grandiose terms. "The Ellenberger is still there, waiting to ruin every well with a torrent of water. But we have learned to respect it, not fear it," he said. Johnson County was the home of "shattered dreams [and] dry holes," Marble said. By early 2004, Marble and Hallwood had learned to tame Johnson County. In January he gave a presentation at the Fort Worth Petroleum Club. The club was up on the thirty-ninth floor of a downtown skyscraper. The windows faced north, toward the area where almost all of the Barnett Shale activity had taken place. The fifty-six geologists and engineers who attended had their backs turned on Johnson County.

Marble shared the results of Hallwood's latest wells to the south.

It had learned to conquer the Barnett, using horizontal wells, even when the rock was above the Ellenberger, he boasted. The room was flabbergasted. "Hallwood flips this data up, and the whole room just said, 'Wow,'" recalled Keith Hutton, a former executive vice president of operations with XTO Energy, the largest company in Fort Worth. The message was clear: the Barnett Shale extended straight through the city and probably included several counties to the west as well. Geologists knew that the rock extended far and wide. Hallwood demonstrated that the industry could make profitable wells across many counties. XTO began to buy up companies for their acreage. Five years later, a member of XTO's board of directors named Jack Randall called up Exxon CEO Rex Tillerson to sound him out on a deal. Randall knew Tillerson from their days together in the University of Texas marching band. Randall played trumpet; Tillerson played drums. They met in Tillerson's office in August 2009. Randall said that XTO was looking for a buyer. "I think we'll be interested," said Tillerson. Three months later, Exxon agreed to buy XTO and its shale assets for $31 billion and assume $10 billion of the smaller company's debt.

In June 2004 Hallwood drilled the Lakeview #1H overlooking Lake Pat Cleburne. It used a massive amount of water to fracture the rock, much more than typical, and produced one of the best wells ever in the Barnett. It roared to life at 6.8 million cubic feet per day. Due to its minority stake, acquired as part of the Canaan deal, Chesapeake owned a piece of this home run. News of this well made it up to Oklahoma City, where McClendon decided it was time to get in on the shale game. Wall Street was talking it up, and Chesapeake's big competitors were acquiring stakes in and around Fort Worth. Later that year, Hallwood put some of its Johnson County acreage up for sale. Chesapeake rushed to be first into the data room and made a bid it hoped would freeze out competitors. But Hallwood thought it could get more and kept the auction going, eventually settling on another company. When that bid fell apart, Hallwood called Chesapeake in late November. The next morning, McClendon flew to Dallas and over breakfast bought the assets for $277 million.

This bite of the Barnett Shale wasn't enough for McClendon. Within a few months, he spent another $250 million to buy more of Hallwood's Barnett assets. If Devon was conservative, McClendon was aggressive. He wanted more of these shale assets, and that meant he needed a lot of money to finance his plans. Chesapeake ended 2004 with $3.1 billion in debt. A year later, Chesapeake's debt rose to $5.5 billion. Then it hit $7.4 billion and $11 billion in subsequent years. As 2008 drew to a close, the company had borrowed $13.2 billion.

While Chesapeake was playing catch-up in the Barnett, other companies were testing just how unique the Barnett really was. Southwestern Energy announced that it had discovered the Fayetteville Shale in Arkansas. Its stock jumped 11 percent. Chesapeake decided to rush into this new shale as well. "Once we saw success there, we understood that the world had changed in our industry and would likely never be the same again," said McClendon.

McClendon sensed that shale gas was a disruptive technology long before Nichols. Devon bought Mitchell, its collection of Barnett acreage, and engineers who had firsthand knowledge of modern fracking techniques, but the company failed to grasp that the shale gas had changed the US energy landscape. Only slowly did Nichols change Devon to adapt to the new world. He had taken over as chairman and chief executive of his father's company in 2000 and didn't want to destroy it on a gamble. McClendon had no such institutional baggage. He had built Chesapeake from scratch. Back in the early 2000s, Chesapeake was a small company with few assets that got little love from Wall Street. He was much quicker to tear down the company and rebuild it to focus on shale exploitation. He had less to lose. There wasn't much to tear down.

Oklahoma City is in many ways a small town with outsized ambitions. Both Nichols and McClendon embody this character. For all their outward similarities, they are very different men. Nichols always seemed to be one step above McClendon. He attended the best private high school in the city; McClendon attended the second best. They both headed east for college. McClendon went to Duke, the Ivy League of the South; Nichols to the actual Ivy League. But these two prominent business leaders have had relatively few business dealings or social interactions with each other. This distance almost collapsed in 2006, when McClendon put together a group of friends and business associates to purchase the Seattle SuperSonics, later moving them to Oklahoma City. McClendon sounded out Nichols as possible partner. Tom Ward, a longtime McClendon business partner at Chesapeake, made the approach. Nichols turned them down.

I met with Nichols in April 2012 and asked if he hadn't been too conservative and missed an opportunity to dominate shale exploration in the United States. "In hindsight, was it a mistake at the time? No, I don't think so," he said. He explained that no matter how good shale gas looked, he was wary of putting too large a bet on it. His vision for Devon was a balanced company, with some gas and some oil, some shale and some conventional. "There's a simple reason," he explained. "I've never met the person smart enough to know when oil and gas are going to go up and down." The contrast with his Chesapeake counterpart couldn't be starker. McClendon placed billion-dollar bets on the direction of natural gas prices.

For a while, both Devon's conservative approach and Chesapeake's more aggressive approach to the business created success. Both companies grew large and prosperous. On the north side of Oklahoma City, McClendon built a sprawling corporate campus that resembled an elite East Coast college. Downtown, Devon moved out of its boxy nineteen-story 1980s-era building. It built a new fifty-two-story metal-and-glass skyscraper that dominated the Oklahoma City skyline and filled it with Devon workers. It was the tallest building between Dallas and Chicago. The main entrance to the tower is a soaring light-filled atrium with a small brass plaque embedded in the floor in the middle. It reads "Integrity."

On the day I met with Nichols, the differences between the companies was obvious. A couple of days earlier, serious questions about McClendon and the company he had created had surfaced. At the root was the question of whether there were any checks and balances at Chesapeake, or was McClendon allowed to push it as fast as he wanted and enrich himself along the way. McClendon had taken up to $1.4 billion in personal loans from a private equity firm—at the same time as he was negotiating with the firm to sell it Chesapeake assets. It was difficult to figure out how McClendon could be looking out for himself and for his shareholders simultaneously. Moreover, the Chesapeake board of directors said that it knew nothing about the loans. Confidence in McClendon and Chesapeake began to evaporate. The stock price plunged.

I asked Nichols about his relationship with McClendon. For the first time in a lengthy interview, he declined to answer my question. I asked whether, if Chesapeake's stock fell any further, would he consider making a bid for the company, snapping up the assets of his rival? "I want to touch that, but I am not going to," he said, before laughing heartily. Then he couldn't resist a small dig. "They are a complicated company," he added. "Hugely complicated."

8

THE RISE OF AUBREY McCLENDON

Americans like our abundant energy, but not the men who provide it.

When all is said and done, Aubrey McClendon is apt to be regarded as a visionary, the chief apostle of an energy revolution that left America richer, cleaner, and freer to pursue a new course in the world. With this in mind, a couple weeks after visiting with Larry Nichols, I returned to Oklahoma City to witness a very American spectacle. I came to see a public rebuke of McClendon.

His journey, which is intertwined with the growth of Chesapeake Energy and fracking, is one of the most fascinating recent stories in the annals of American business. It is replete with opportunism, insight, risk, and breathtaking materialism. Understanding how the energy boom unfolded requires understanding the person whose personality most influenced it.

My destination on a beautiful June day in 2012 was Chesapeake's headquarters, which doesn't look like any other oil company I've ever seen. There is no glass-sheathed skyscraper atop underground parking. Instead, when I arrived at the campus, there was a muscular instructor leading an aerobics class in the midst of a campus of low-slung buildings with tree-lined pathways and a creek. A group of young employees on a grassy field played kickball within sight of Aubrey McClendon's office. It is easy to feel like you've taken a wrong turn and ended up at the Silicon Valley headquarters of a high-tech firm.

I was there for the company's annual shareholders' meeting. A year earlier, the two reporters who attended the meeting were allowed to sit with shareholders and talk to McClendon afterward about his vision for energy and Chesapeake. This year, however, there would be no audience with McClendon. Reporters who attended were held in quarantine, allowed to watch the proceedings on a closed-circuit television feed. After I parked my car, a Chesapeake employee escorted me into an underground conference room. The unending blue sky was visible, barely, from windows that looked out onto shafts cut into the earth.

Before they took us to the subterranean bunker, a smiling Chesapeake official asked me to sign a form that said I would be removed from the property if I attempted to wander about and interview shareholders or employees. I would not be allowed to grab lunch at one of the company's three on-site restaurants, which a video screen advertised were serving pan-seared Pacific red snapper in a blood-orange beurre blanc sauce with roasted Japanese eggplant.

McClendon was under siege, his leadership and personal finances attacked by activist shareholders. They believed that the assets he had built, leases to drill on millions of acres above US shales and trillions of cubic feet of natural gas, were worth more in the hands of someone else. He had relentlessly—and, some argued, recklessly—created the preeminent shale energy company in the world. Now his hold on the corner suite was tenuous.

Shortly after the meeting began at ten o'clock, McClendon took the stage to applause. He wore a black suit, a white shirt, and a pink tie. He stood at a podium, on a green carpet, dwarfed by a neon green screen. He didn't acknowledge the applause in the room.

In the world of American capitalism, the previous few weeks had resembled an "Oklahoma Spring": a populist revolt against a man with absolute control over a major energy company. In theory, he answered to a board of directors, but in practice, he often operated outside the view of these men and women. McClendon had cofounded the company and was its guiding spirit. He exercised a remarkable level of control, weighing in on small land deals and even writing quotes to be put in the mouths of senior executives. One of the items on the meeting agenda was the vote to reelect two members of this board. Shareholders from near and far could give voice to their displeasure with McClendon's stewardship by voting against the corporate overseers.

Twenty minutes after the meeting started, corporate secretary Jennifer Grigsby came to the podium to announce the preliminary results of voting on the two board members who had stood for re-election. One was Richard Davidson, a former chairman and CEO of Union Pacific railroad. He had served on the board for six years. The other was V. Burns Hargis, a lawyer, bank executive, and president of Oklahoma State University. He had been a director since 2008. Both sat on the audit committee, charged with overseeing Chesapeake's finances. The Sarbanes-Oxley Act of 2002 required a strong, independent audit committee, one of the law's key provisions. If anyone should have been keeping an eye on McClendon's complex web of loans and finances, it was Hargis and Davidson.

Grigsby read the tallies without a trace of emotion. Davidson had received support from only 27 percent of votes cast. Hargis had done slightly worse: 26 percent. The vote of no confidence was historic. A group that tracks corporate governance said that Hargis's 26 percent was the lowest level of support for a board member of any of the five hundred largest public American companies in at least five years—and probably longer. Under the topsy-turvy rules of corporate democracy, it was enough to be reelected. But both men turned in their resignations anyway.

McClendon took the stage again and launched into a defense of his accomplishments. The shale revolution had begat nothing less than "a reinvention of the US, and, someday, the world's energy business . . . it will be one of the most important developments in the world as we move away from an economy that has been under the stranglehold of OPEC." He promised a new Chesapeake, a pledge he had made before but never quite fulfilled. The new Chesapeake, he went on, would be "simpler . . . we have what we own, and we are happy with what we own."

At the end of his short remarks, he mentioned the public drubbing his handpicked board members had just received, saying, "We will be studying the results of the vote today and see what else needs to be done." After he spoke, Vincent Intrieri took the microphone. He worked for Carl Icahn, a well-known activist investor with a reputation for acquiring shares in a company and forcing change. Icahn had been buying Chesapeake shares for several weeks. Intrieri was in his midfifties. He wore an expensive suit and eyeglasses, the uniform of a powerful capitalist. "We believe Aubrey that you are a great oil and gas man," he said, "but even great leaders need oversight." The new board of directors, he said, should keep an open mind and consider a potential sale of the entire company.

After barely an hour, the meeting wrapped, and the video feed into the reporter's bunker was cut. I prepared to file to my *Wall Street Journal* editors an early version of my account of the vote and McClendon's public rebuke. The speakers, which had piped in the sound of the meeting, began to play music. It was the unmistakable voice of Billie Holiday, singing in her melancholy voice about the brutality and fickleness of love.

I shook my head and chuckled at the bizarre coincidence. The song, "Fine and Mellow," provided a strange commentary on how shareholders had fallen out of love with McClendon's and Chesapeake's story. But were shareholders the mistreated lover in Holiday's song? Or was it McClendon himself, loved and then discarded by the shareholders?

As I sat there listening to the song, I wondered how we had arrived at this point. In the decade since Devon had purchased Mitchell Energy, McClendon and his pursuit of American oil and gas had changed the energy world. President after president, from Nixon onward, had pledged to do all they could to set the United States on a path toward energy independence. All had failed. McClendon and Chesapeake had done as much toward that goal as all of them put together, but there was no invitation forthcoming to a White House ceremony honoring his accomplishments. Indeed, if you define energy independence as reaching the point at which the US economy and foreign policy could no longer be manipulated by countries that sell us energy, we were rapidly approaching that long-sought goal.

But there was little reason to celebrate on Chesapeake's campus. The company was selling off assets to pay its debt and faced a cash shortage. Investor confidence was shaken. Its bonds, already listed in junk status, had been lowered even further. To staunch the financial bleeding, Chesapeake had taken an emergency $3 billion loan at a steep interest rate. The loan had come from Wall Street giant Goldman Sachs and Jefferies Group, an investment bank with deep ties with Chesapeake. Jefferies was not only Chesapeake's go-to banker and fund-raiser for the better part of a decade, but one of its stars was Ralph Eads III, McClendon's longtime friend and godfather to one of his children.

"Yes, it's a crisis," wrote Jon Wolff, an energy analyst who had followed Chesapeake for years, in a note to his clients. A couple days after the meeting, a photograph of the second game of the 2012 NBA finals between the Miami Heat and Oklahoma City Thunder, began making the rounds of Wall Street. Sitting courtside in a Smurf blue "Team Is Family" Thunder T-shirt, was Fu Chengyu, the acquisitive chairman of Chinese oil company Sinopec Corp. Energy gossip circles lit up. What was he doing in Oklahoma City? The seat was courtesy of McClendon, a Thunder co-owner, everyone assumed. Was McClendon, the leader of the shale revolution, about to sell out to the Chinese?

How had McClendon, the most influential energy visionary in a generation, reached this point of disgrace and doubt? The seeds of this downfall were planted years earlier when Chesapeake struggled in its first years as a public company. Looking back, one could see McClendon's ambition, arrogance, and hubris, traits that would ultimately stain his career.

What is also visible in the story of his rise to prominence and his fall from favor is a larger truth about the energy boom. Once engineers demonstrated that shale rocks could be fracked and exploited, a race among energy companies began. It would be led by executives with a talent for promoting their companies and raising enormous sums to keep drilling. In this, McClendon was without peer. To understand the rise of fracking, it is necessary to understand how

McClendon, a financial engineer, came to dominate an industry once led by petroleum engineers.

Aubrey Kerr McClendon was born in Oklahoma City in 1959. His middle name marked his family ties to the oil business and to politics. His grandfather's brother, Robert Kerr, helped found Kerr-McGee, the pioneering oil and gas company that drilled the first offshore well in the Gulf of Mexico. Kerr used his business success to launch a political career, serving as Oklahoma governor and later in the US Senate. A campaign slogan from a few years before McClendon was born mixed populism with braggadocio: "I'm just like you, only I struck oil."

Aubrey's father, Joe C. McClendon, worked for decades at Kerr-McGee. A major employer in Oklahoma City, its portfolio ranged beyond drilling to include mining uranium and manufacturing plutonium pellets for nuclear power plants. A young woman named Karen Silkwood, subject of the 1983 movie for which Meryl Streep was nominated for an Academy Award about her whistle-blowing and mysterious death, worked at a Kerr-McGee subsidiary in the early 1970s. Joe McClendon, however, wasn't involved in either cutting-edge fossil-fuel exploration or the dangerous world of atom splitting. He sold gasoline and other petroleum products to gas stations. When the younger McClendon spent time with his father on the job, there was no romance of the oil rig. "I spent time looking at dirty bathrooms in gas stations," he remembered years later.

He attended Heritage Hall, a relatively new private high school that offered a less rigid social environment than Oklahoma City's other elite prep schools. Aubrey excelled both socially and academically. He was elected senior class president and was covaledictorian. Graduating in 1977, he went off to Duke University. His years in

Durham, North Carolina, revolved around schoolwork, the hard-partying Sigma Alpha Epsilon fraternity, and the school's basketball team. One of his best friends was Ralph Eads, a fellow fraternity member. Eads hailed from Houston, and his father, like McClendon's, worked in the oil industry. They lived across the hall from each other. Eads was known as Ringo. People close to McClendon called him Aubs. Ringo and Aubs.

During their senior year, McClendon was wrapping up a degree in history, but he had also taken a number of courses in economics and finance. Eads studied economics with an eye toward a job on Wall Street. For the final home basketball game of their four years in college, Eads and McClendon must have arrived early at Cameron Indoor Stadium on the morning of Saturday, February 28, 1981. When the doors opened, they grabbed seats practically on the court, almost immediately behind the players' bench. It was as close to the action as possible.

The game was one last chance to see their beloved Blue Devils as students. The opponent was the school's archrival, the University of North Carolina Tar Heels. The Blue Devils had sputtered that year under first-year coach Mike Krzyzewski, including two losses at the hands of the Tar Heels. Under legendary coach Dean Smith, UNC was ranked eleventh in the nation. It was the establishment behemoth. Duke was at risk of faded glory, a once-proud basketball program that had fallen into disrepair.

Duke kept the game close and trailed by only two points with two seconds left. After a time out, senior forward Gene "Tinkerbell" Banks caught the inbound pass at the top of the key, turned, and shot the ball over his defenders' outstretched hands as time expired. The ball went straight through the rim, tying the game and sending it into overtime. Banks was electric in the extra period, notching several rebounds and baskets. In the final seconds, when a teammate

missed a shot, Banks grabbed the rebound and laid it in for the victory. Thousands of Duke students rushed onto the court in euphoric
celebration. These days Duke is a basketball dynasty. That day, a
David had defeated a Goliath. Eads recalls being among the throng
that celebrated on the hardwood. McClendon has a different mem-
ory. He says both of them were in a rush to drive an hour west to
Greensboro, where they had tickets to a Bruce Springsteen concert.

By an odd coincidence, I was among the ten thousand people at
Cameron for the game. My family had moved to nearby Chapel Hill
for a year, and my father took me to the game that day. We sat about
as far from courtside as possible, just a couple rows from the rafters.
When the game was over, we got in our car and drove ten miles back
home. On Monday I went back to fourth grade and told my friends
about having been at the game. A few months later, at the end of the
school year, my family moved back to Philadelphia.

McClendon moved home to Oklahoma City a few weeks later
after the basketball game, finishing his time at Duke with a 3.7 grade-
point average and a fiancée. Later that summer, he married Kathleen
Upton Byrns, a granddaughter of the founder of Whirlpool Corporation. Eads, who was a member of the wedding party, went to New
York and began training as an investment banker at Merrill Lynch,
where he learned how money moved through the modern financial
world. Years later, Eads and McClendon would work together again,
the financier finding money for McClendon to keep up a torrid
pace of leasing and drilling. I was again orbiting around the action
as a *Wall Street Journal* reporter keeping tabs on Chesapeake. I first
interviewed McClendon in August 2006 in Oklahoma City. On my
flight home, I wrote in my notebook: "Is he a huckster, a dynamic
salesman, a visionary, a fool? I can't tell."

A couple months before McClendon returned home from Duke, oil prices hit $39 a barrel. Revolutions and war between Iran and Iraq drove up crude oil to this never-before-seen level. Adjusting for inflation, this 1981 price was the equivalent of $100 oil. A generation would pass before crude regained this plateau. As McClendon settled in Oklahoma City, crude prices were falling due to a global recession and increased output from the Middle East. Drilling rigs idled. And aggressive loans to oil companies began to cause trouble at the Penn Square Bank, a major financial institution in Oklahoma City where McClendon, as a kid, had deposited money from mowing lawns and selling holiday cards door to door. The bank had been cofounded by McClendon's great-uncle Robert S. Kerr. Bank auditors found Penn Square had too many loans that weren't being repaid, there was not enough cash, and inexperienced loan officers had too much autonomy. The bank was awash in loans to oil companies. When oil prices dropped, these loans began to go unpaid. Penn Square had farmed out its loans to other banks across the country. When Penn Square went under, its loans sunk the Continental Illinois National Bank in Chicago, then the seventh largest in the country. The slow-motion collapse of Penn Square turned Oklahoma City from a boomtown into a ghost town. The regional economy of the US oil patch sunk into the doldrums.

In the midst of this crisis, young Aubrey McClendon needed a job. The small company where he landed, Jaytex Oil & Gas, was owned by his uncle, Aubrey Kerr Jr. His uncle offered him a position as a staff accountant. It was steady work but, like his father's job servicing gas stations, it was not glamorous. After nine months at Jaytex, McClendon took on a new assignment in the land department, where he was responsible for figuring out who owned the mineral rights to any particular parcel and getting them to sign a lease. Mineral rights are often split, and landmen often must hunt down a long-forgotten family member who owns one thirty-second of the oil and gas. The work involves a lot of digging through county court-house land records. This work appealed to McClendon. "In land, I found my true love in the oil/gas business," he wrote in an email to me in 2006. "Learning to become an oil/gas landman combined my love of history with my appreciation for the precision of numbers learned through accounting."

In 1982 McClendon realized that Jaytex wouldn't survive the downturn. He left the company and struck out on his own. He bought a typewriter, some oil-and-gas maps, and rented a one-man office. He spent the next few months acquiring leases and trading them. He soon began to bump into another independent landman named Tom Ward. If McClendon had been born into a privileged life, Ward was the opposite. He came from Seiling, a tiny town in northwestern Oklahoma, into what he called "a fairly dysfunctional family." Ward can be gruff, while McClendon is all polish and charm. Ward went to the University of Oklahoma, a far cry from elite Duke. After graduating, he struck out on his own as a landman. His early success in the business came not from geological acumen but from recognizing and exploiting Oklahoma's oil and gas laws.

In southern Oklahoma, there is a large gas field called the Golden Trend. A company that wanted to drill a well might acquire the mineral rights to the vast majority of a 640-acre block of prop-erty but run into problems locking up all the rights. In this situation, the company could apply to the state for a forced pooling order. If granted, all of the mineral owners in the block who had not signed a lease would be forced to participate, or pooled. The farmers (or whoever owned the mineral rights) who hadn't signed would be entitled to the highest amount paid to the farmers who had signed. Ward figured that if a big energy company wanted to go through the bother of obtaining a forced pooling order, it must be excited about the geology and prospects. He spent his days driving around and locating the holdouts, explaining that they were going to be pooled and then offering them a bit more than the state would pay. In this way, he would get a small slice of another company's well—without the cost of hiring geologists and engineers to do any actual exploration work. He was, in essence, piggybacking off the work of other companies. This strategy proved successful, except that he kept finding himself in competition with McClendon, who had figured out the same trick. They bid against each other, driving up prices and eroding each other's profits. In the summer of 1983, McClendon and Ward agreed to collaborate instead of compete. This partnership was the beginning of Chesapeake Energy.

From that point on, the two men often pursued separate deals but worked under a fifty-fifty agreement that intertwined their interests and split the risks. Any deal that Ward worked out, McClendon would take a 50 percent share—and vice versa. Despite joining forces, the partners maintained separate offices in different buildings. Indeed, in the twenty-three years they worked together, they never had offices in the same building.

Neither man any had training to read a well log or study subsurface geology, the typical background for oil company executives. What they knew was land and money. Recognizing their limits, they weren't attracted to the high-risk world of drilling wildcat wells. "We clearly could not outthink a geologist or an engineer," McClendon said a few years ago. So the partners pursued oil fields "where once the geology was recognized and the engineering solution had been crafted, that it was the land guys that then made the difference." Success for the young Oklahomans didn't require any particular technology or engineering skill. It required hustle and money, and an ability to lease land before anyone else.

By the end of the decade, the partnership had outgrown its

model of buying small slices of other people's wells. Ward and McClendon migrated to acquiring large tracts of land, drilling and operating their own wells. Ward gravitated toward operations. McClendon took over finance and land operations. In May 1989 the partners incorporated Chesapeake Operating. Even after the early-eighties oil bust, Oklahoma remained filled with hundreds of small-time operators, raising money and drilling a handful of wells. Even today, there is one oil operator for every 1,200 Oklahomans, nearly twice the rate of Texas. Getting into the business wasn't hard. It took the gift of gab and the ability to raise money from friends, family, and business associates. "Oil is *the* Oklahoma business, and it's a family business," said Dewey Barlett Jr., part of a family involved in both energy and Tulsa politics. It was a point of pride for many Oklahomans to be invested in a well or two. These deals were free-wheeling, often sealed with a firm handshake, a confident smile, and the promise that everyone was putting his own money into the venture. This arrangement worked just fine, except when the foundation of trust crumbled. Then acrimony and lawsuits took over.

That's what happened in an energy deal that turned into a lawsuit which has been largely forgotten, and misunderstood by those who remember it. The case involved allegations of fraudulent land deals, double-crossing petroleum engineers, and what turned out to be a dry hole in rural Beckham County. Tom Ward had put together what he called the East Virgil prospect: land in westernmost Oklahoma, near the Texas panhandle and less than a mile from a successful well. He hired a geologist to prepare a subsurface map of his acreage that suggested there might be oil and gas there.

Ward had leased the land and put together the package but was looking for someone else to take over and drill the prospect. (As usual, McClendon and Ward split the prospect evenly.) Ward offered the deal to Ralph E. Plotner Oil & Gas, a company owned solely by

Ralph Plotner, an Oklahoma City salesman for a local radio station who had decided to get into the oil business a year earlier. On his first venture into his new trade, the neophyte had trudged through snow to a farmhouse, got the farmer to sign a lease, and drilled a successful well with the help of a friend. Plotner Oil agreed to buy a stake in the East Virgil prospect and took over as the operator. Then Ward rounded up another investor, Continental Trend Resources, run by Harold Hamm. (The company is now known as Continental Resources and is a major oil producer; Hamm is regularly listed as one of the country's wealthiest billionaires.) Eventually Plotner Oil bought a 40 percent interest, Continental took 20 percent, and another small company held 25 percent. The remaining 15 percent was held equally by TLW Investments, a company wholly owned by Ward, and Chesapeake Investments, a company owned by McClendon and his wife.

After the well was drilled and didn't turn up any oil or gas, the partnership soured. It all ended up in a lawsuit in which Plotner Oil claimed that Ward had lied to him while promoting the East Virgil prospect. He said he was impressed when told that Ward and McClendon had had personally invested more than $2 million in the prospect, paying $500 an acre to accumulate between 4,500 and 5,000 acres. But during the trial, it came to light that they hadn't spent $500 an acre. They had spent only $300 an acre, according to testimony that the jury relied on.

Megan Hann, a petroleum geologist who worked for McClendon and Ward, was subpoenaed to appear at the trial. Her testimony damaged her employers. She said that Ward regularly overstated reserves, or recoverable oil and gas, to potential investors. At first, she said, the embellishments were small, but by 1990 "the exaggerations got pretty large." Hann said she sat in on meetings where she heard Ward make these outlandish claims, but remained silent. She attended these meetings to present the geology, not the economics. "I didn't believe it was my place" to contradict Ward when talking to potential investors, she said. But when her friends and family asked for tips, she testified that she advised them not to put in any money. She also said that she heard Ward tell Plotner he had spent $500 an acre, and repeat the same to Harold Hamm.

One day, driving back from Hamm's office in Enid, she asked Ward why he had spent so much on what was basically an unproved prospect. "I was surprised that they had bought the acreage for five hundred dollars an acre," she testified under oath, "and he said that they really had not bought it for five hundred an acre. It was—they had three hundred dollars—around three hundred an acre into the prospect." Harold Hamm also testified that Ward had told him he bought the acreage for $500 an acre, and that he later learned that Ward had actually paid less. This testimony would prove crucial. The jury believed that Ward and McClendon had misled Plotner and other investors in the East Virgil prospect. It handed up a verdict against Ward and McClendon, awarding Plotner Oil $904,000 in actual damages and $1.25 million in punitive damages. Ward and McClendon appealed, but the appellate court sided with Plotner Oil and upheld the award. The Oklahoma Supreme Court let the lower court ruling stand.

The appellate court also discussed a damning piece of evidence about the founding of Chesapeake Energy. Plotner had testified during the trial that he had relied on a consulting petroleum engineer named Kenny Davidson, whom he had hired to evaluate the prospect. The justices wrote that "sufficient evidence on the record existed to support a conclusion by the jury that Davidson was secretly working for Ward and McClendon to defraud Plotner Oil even before he was officially hired as the first employee of Chesapeake Operating." Davidson remained at Chesapeake until he retired in 2005.

A couple days before the Oklahoma jury began deliberating the case, Chesapeake Energy filed paperwork for an initial public offering (IPO). Chesapeake Operating had grown quickly as it got into the business of drilling its own wells. In 1992 the company sold oil and gas worth $10.5 million, up from less than $400,000 two years earlier. The number of wells it owned went from two to twenty-nine. Chesapeake Energy had been formed only a couple years earlier, pulling together assets held by Chesapeake Operating and other entities owned by Ward, as well as by McClendon and his wife.

The new company struggled to find money to keep its doors open. The young executives tried to get traditional bank loans but were turned away. When they did find lenders, the deals were on onerous terms. McClendon secured a line of credit from the Trust Company of the West (TCW), a large Los Angeles investment group, and drew it down quickly. Then Chesapeake secured another loan from Belco Oil & Gas, an energy business owned by the wealthy Belfer family in New York City. An earlier family-backed company, Belfer Petroleum, had merged with Houston Natural Gas, a company that formed the basis of Enron. A member of the family had remained on the Enron board until the end and lost billions of dollars on paper when the stock became worthless. To get both loans, which carried a stiff 9 percent interest rate, both Ward and McClendon had to guarantee the loans with their own personal holdings.

Chesapeake's financial position was precarious. A few months before filing for an IPO, the giant firm Arthur Andersen resigned as Chesapeake's independent accountant. One of the most important duties of a public company accountant is to tell investors whether it believes the company is at threat of financial liquidation over coming months. Arthur Andersen wasn't confident that Chesapeake met that "going concern" definition. It wouldn't give its imprimatur. Chesapeake disagreed and found another accountant. A few years later, Arthur Andersen, one of the "Big Five" accounting firms, surrendered its license in the wake of the Enron implosion. It had been Enron's auditor, and hadn't raised flags about it as a going concern.

McClendon was the chairman and chief executive of the new company, Ward the chief financial officer. Both men were paid $175,000 in salary, but were given an unusual perk. Each could purchase a small 2.5 percent stake in every well that Chesapeake drilled. To prevent cherry picking, they had to choose each year whether to invest in all the wells the company drilled or decline to participate altogether. Ward and McClendon would pay their share of the well costs and receive proceeds from any oil or gas produced. Over the next year, Chesapeake spent nearly $40 million drilling wells. McClendon and Ward's share of the costs was $1 million each—several times their annual salary. It was an unusual setup that required the executives to borrow large sums. Years later it would cripple McClendon's career and become a major reason why shareholders rejected its two board of directors candidates in 2012.

Overseeing the perk was a board of directors that wasn't exactly a paragon of independence. In addition to McClendon and Ward, there was a childhood friend of McClendon's who did legal work for the company. McClendon and Ward were in hock to the other four directors. In May 1992 these four lent $1.65 million to the company's founders, in an unusual financial arrangement to help McClendon and Ward pay off money they owed the company. In the post-Enron financial reforms instituted a decade later, this kind of loan was prohibited due to the conflicts created. The directors owed their paid positions to McClendon and Ward. The executives' pay and perks were approved by the directors. If the directors felt it was necessary to fire one of the founders, wouldn't it be human nature to hesitate, as McClendon and Ward might not be able to repay their personal debts to the directors? It was a knot of conflicts that left shareholders on the outside.

Chesapeake was the worst performing IPO of the year, McClendon said in a webcast with investors about a decade later. A share of the company ended 1993 worth less than half what it sold for on the first day of trading on Nasdaq. But the money raised selling shares to the public allowed the company to pay down some of its debt to TCW, and it soon secured a more conventional bank loan. Even after raising about $25 million in the IPO, McClendon prowled for more money. He was willing to pursue unconventional sources of funds, inventing them if necessary. In October of that year, he struck a deal with another energy company to help cover the costs of developing a gas field in Texas. Chesapeake boasted that it "believes a financing arrangement of this type is unprecedented in the industry." This creativity helped keep Chesapeake with enough money to keep drilling. Soon Wall Street began to notice this small company and its gutsy chief executive. From the beginning of 1994, the stock soared for two years, rising nearly 2,000 percent. McClendon drove Chesapeake like a fancy race car, speeding through the curves.

In the first few years, Ward and McClendon worked around the clock. Ward generally arrived at the office between five and six in the morning, put in twelve hours, ate dinner at home, and then worked more. McClendon came in to work later but would often stay at work until two o'clock or later. "There were days when I would be coming in, and he would be going home," said Ward. "We believed we could overcome any deficiencies with hard work." Ward's job was to find places to drill and let Aubrey figure out how to get the money together to finance the growth.

McClendon's goal was to become a big company. But even as Chesapeake grew, Ward insisted on knowing about every well and often visited the field to talk to the drillers and crews. When he left the company, in 2006, it was because Chesapeake had grown so large that he couldn't be hands on anymore. "It was causing me a lot of stress," he said. "I couldn't keep up." I asked him if he ever considered asking McClendon to slow down, during the early years or later on. He chuckled at the suggestion. "Oh, no. That wasn't our model. We were only going to get bigger. That was our lifestyle. That was Chesapeake." I asked how he would describe the company he had helped create. "More. More," he answered.

Within a couple years, a flaw emerged in Chesapeake's business model. To continue growing its production and profit, as Wall Street wanted, it drilled a lot of wells. Soon the company's cupboard of drilling opportunities ran low. It needed to restock to continue its torrid pace. An opportunity presented itself in late 1994, when Occidental Petroleum drilled a successful deep well into a Cretaceous-era rock formation called the Austin Chalk in central Louisiana. The Los Angeles company was so excited about the well that it put it on the cover of its annual report. Chesapeake was making good wells in the Austin Chalk a couple hundred miles to the west in Texas, using similar techniques. McClendon decided to follow Occidental's lead, despite signs that Louisiana wouldn't prove as easy as Texas. Wells on the eastern side of the Sabine River were deeper and under more pressure, which meant that they were more expensive. And the Louisiana wells produced more water along with oil, raising costs further. What's more, Chesapeake wasn't entitled to the kind tax exemptions in Louisiana that it received in Texas. Facing this challenged drilling environment, Occidental leased one hundred thousand acres over a couple years. Chesapeake wasn't willing to take it slowly. It spent

$179 million vacuuming up more than one million acres, an area only slightly smaller than Delaware.

Chesapeake said this new Louisiana acreage would be the "focus of the Company's exploration and development activities in the foreseeable future." It budgeted $125 million for drilling in 1997 and borrowed $200 million to fund its efforts. By early 1997, Chesapeake told investors it had hundreds of well locations waiting to be drilled, more than twice as many as its nearest competitor. There was just one problem. The wells produced oil and gas, but not enough to justify the cost of drilling them. Chesapeake had gambled that the geology it was accustomed to in Texas would be uniform across a wide area. That wasn't the case. In early 1997 it spent $40 million to drill ten wells. The company generated barely $1 million in oil and gas for its effort. A couple of Wall Street analysts began raising questions, including one who noted that Chesapeake's estimates for how much oil and gas could be found in Louisiana "seem quite aggressive." In June, after months of positive reports about the Louisiana Austin Chalk, Chesapeake precipitously changed its message, stunning investors. Most of the acreage it had leased wouldn't support profitable wells. It would write down the value of its investment in Louisiana, and it wouldn't grow as quickly as it had previously promised. The market clobbered its stock on the day of this announcement, and the share value continued to plunge in the following months. By July 1998, seeing few viable possibilities, the board of directors put the company up for sale. The stock was worth one-third of its value the day before announcing the abysmal well results in Louisiana. There were no serious offers for Chesapeake.

McClendon has always been a voracious consumer of information. Ralph Eads recalled how he was impressed that in college, McClendon's dorm room was full of magazines and books that weren't on the curriculum. He was just reading them. Years later, when Chesapeake built a corporate gym, a plastic file holder was attached to the wall next to McClendon's preferred workout machine. On the holder was written, "Aubrey's Reading File—Please Do Not Disturb." His staff made sure it was full by the time he arrived in the morning for his exercise. McClendon's urge to gather more information than anyone else led him to seek a meeting in early 2000 with Peter Cartwright, founder and chief executive of Calpine, a California company that owned power plants that used natural gas to generate electricity. McClendon and his chief financial officer, Marcus Rowland, hopped on the company's corporate jet, a seven-seat Cessna Citation II, and flew out to San Jose for a lunch meeting at a restaurant near Calpine's headquarters. Calpine was an aggressive company in the recently deregulated power market. It owned power plants with a capacity to generate 4,273 megawatts, and was either building, or planned to build, enough new plants to quadruple its power capacity.

McClendon crunched the numbers and realized that if Calpine did build all those new power plants, it would burn through five billion cubic feet of natural gas a day, or about a 10 percent increase in US gas consumption. Calpine's view was that gas was inexpensive, and companies such as Chesapeake could find more. But over the previous few months, McClendon had grown convinced that the American energy industry was going to struggle to keep gas production steady, much less increase it. If demand was headed up—and supply wasn't—the gas market was about to change. After lunch, the Chesapeake executives headed back to the airport. As they boarded their jet to return to Oklahoma City, McClendon turned to Rowland and said, "We got a chance."

His insight was that natural gas prices would increase. "I knew that all I needed to do was go long on natural gas and we would deliver one of the best comebacks in the history of the business," McClendon later said. Rising prices meant it made sense to buy other gas producers as soon as possible, and justified drilling the deeper, more expensive wells that were Chesapeake's specialty. McClendon's hunch soon proved correct. For the previous few years, natural gas had traded in a narrow band between $2 million and $3 per million British thermal units. But the investments in drilling had fallen, and in late 2000, cold winter weather over most of the United States led millions of people to turn up their thermostats. November 2000 was the coldest in decades, and temperatures in December were below average in forty states. Demand for gas spiked, and so did prices. The price of gas topped $10 across the country and went significantly higher on the West Coast. Consumers who opened their December home heating bills were in for a nasty surprise. Big industrial consumers, such as fertilizer, steel, and petrochemical manufacturers, who bought gas on the open market, were pummeled. Chesapeake, on the other hand, did quite well. It generated enough cash to alleviate some of the crushing debt left behind by its misadventure in Louisiana. Its profits for the six months that winter were up tenfold from the previous winter. Even excluding a one-charge accounting gain, its profits from oil and gas soared 400 percent.

McClendon has told this story of Chesapeake's survival many times. It is the period when "phase one" of Chesapeake ends and "phase two" begins. He pieces together a complicated supply-and-demand puzzle before nearly anyone else and acts on it. He becomes bullish on natural gas, thanks in part to his meeting with Calpine, and begins to buy as much gas in the ground as possible.

But there was another reason that gas prices soared in late 2000 and early 2001, helping resurrect Chesapeake's finances. And it had nothing to do with supply, demand, or cold weather. It was market manipulation by a group of energy traders that left California with exorbitant gas prices and rolling electricity blackouts. And one of the key players at the heart of what became known as the California energy crisis was McClendon's best friend: Ralph Eads.

By July 1999, Eads had spent most of his career in investment banking, with a couple unsuccessful forays into drilling wells. In the mid-1980s, Eads had left Merrill Lynch to become chief financial officer of a new Houston company called American Energy Operations. He was not yet thirty years old. With expensive offices on a top floor of the One Shell Plaza skyscraper in downtown Houston, the company had ambitions to match the views. The president who hired him was future Chesapeake CFO Marcus Rowland. In 1986 Eads boasted of the company's abilities, saying, "We believe we can manage several hundred million dollars of assets with very few people." American Energy Operations did about as well as an oil exploration company run by people without any technical expertise would be expected to do. According to Texas state records, it drilled five wells during its three-year existence. Four of the wells were dry. The fifth, in westernmost Texas, produced a paltry thirty-eight thousand barrels of oil over a quarter century. After this, Eads went back to investment banking until he heard the call of a new siren song: the profitable word of energy trading that Enron pioneered and many energy companies wanted to copy.

In July 1999 Eads was recruited to be one of the top executives at El Paso, a staid pipeline company in the business of moving gas around the country. El Paso wanted to reinvent itself in the image of Enron, its flashy neighbor in downtown Houston's thicket of skyscrapers. Enron had undergone a metamorphosis from Houston Natural Gas, a midsized collection of pipelines, into an energy trading powerhouse. As energy markets deregulated and were freed

from government oversight, Enron built up the new business of buying and selling energy contracts. In 1990, gas contracts began trading on the New York Mercantile Exchange. Before this change, gas
was exchanged as a physical commodity. The advent of gas futures
meant gas could be traded as a piece of paper—a financial contract.
Often the energy traders weren't interested in actually taking delivery of the gas they traded. They were interested only in the profits
that could be derived from betting on the price of gas. Engineers had
built the energy industry, but in the 1990s, Houston energy traders
who sat at their desks in front of multiple computer screens became,
for a time, kings.

El Paso watched the adulation Wall Street heaped on Enron and
the profits its traders generated. It wanted to capture some of that
lightning. That was Eads's mandate when he took over the company's new trading unit known as El Paso Merchant. Under Eads, El
Paso Merchant expanded. It bought power plants in Rhode Island
and California. It amassed gas and power contracts, bundled them,
and resold them to large wholesale customers in forty-eight states.
It also provided risk-management services to energy consumers.
Eads seemed to have a golden touch. In the first full year that El
Paso Merchant reported to Eads, the segment reported a $433 mil-
lion operating profit. A year earlier, it had racked up a $65 million
loss.

El Paso owned most of the largest pipelines that connected natural gas from New Mexico's San Juan Basin and Texas's Permian Basin
to the Arizona-California border, offering space on the pipelines to
the highest bidders. In February 2000 Eads pitched El Paso CEO Bill
Wise on the advantages of allowing El Paso Merchant to purchase
the capacity. The company would have "more control of total physical markets," could "help manipulate physical spreads," and would
have the "ability to influence the physical market" to help its traders

profit. The stunning memo lays out plans to control the flow of gas into California to make money. Later, when asked what was meant by terms such as "manipulate," Eads told government investigators the word didn't mean what it appeared to say. He suggested that the language was misleading. Federal investigators disagreed. The words' meaning was simple enough, they concluded. El Paso wanted to choke off enough of the steel pipe highway into California that prices would shoot up. Wise gave his approval, and El Paso Merchant bid successfully for the pipeline capacity. Traders soon controlled enough space in the pipeline to deliver more than 1.2 billion cubic feet of gas every day into California, about one-fifth of the gas used in the Golden State.

Even though they were unaware of the internal El Paso memo, California officials soon grew concerned and protested to the federal energy overseer that El Paso Merchant was in a position to manipulate the market. It filed a complaint in April, a month after El Paso Merchant traders started acting effectively as the gatekeeper for gas headed to California. It would take years for this complaint to work its way through the regulatory process. In the meantime, El Paso traders began to do what the state feared: control gas supply and drive up market prices. The California energy meltdown of 2000 and 2001 is remembered largely as a crisis in the electricity market. Wholesale power prices spiked. There were more than one hundred electrical emergencies declared and a handful of rolling blackouts. But the power crisis was intertwined with problems in the natural gas market. In March 2000, when El Paso Merchant took control of the pipeline contracts, gas at the California border cost $2.84 per thousand cubic feet. By December, the price had risen nearly tenfold, to $25.08. Briefly, prices spiked as high as $60. When prices neared their peak, in November, El Paso Merchant turned up the flow of gas into the state. The Houston traders created a scarcity, watched prices rise, and then profited. They "cashed in big-time on the exorbitant prices," charged the state's lawyer.

California's case against El Paso hinged on Eads's presentation to Wise in February 2000, about the time that McClendon flew out to California to talk to Calpine. Eads was called to testify about his trades in May 2001 in a courtroom in the Federal Energy Regulatory Commission's office in Washington, DC. The judge was Curtis Wagner, a seventy-two-year-old Kentucky native who retained a southern drawl. He owned a thirty-two-foot sailboat named *Hizzoner*. Known as a courtly gentleman, he was generally conciliatory with witnesses. But when Eads evaded Wagner's repeated questions about this presentation, the gentleman judge became angry. "I felt that the witness was not being really open with me. I was having a difficult time getting an answer, so I really lit into him," he recalled.

"I feel you're trying to pull something over my eyes, which I don't appreciate," Wagner told Eads, according to the transcript of the hearing. Eads relented and agreed that he had gone to the CEO and received approval for the deal. "Why didn't you tell me that this morning, and we would have all been home by now?" said the exasperated judge. "You have to get my blood pressure up to get the truth out of you." A power company lawyer in the courtroom said it was like a *Perry Mason* moment. Another lawyer, who represented California, said that Eads came across as untrustworthy. "He was under penalty of perjury, and he was still saying things that were inconsistent with the documents we had, to try to explain away things even though the words were real clear to us," said Harvey Morris.

Having pried the truth from Eads, Wagner concluded that El Paso had manipulated the California gas market, sending both gas and electricity prices soaring. El Paso's stock price plunged 36 percent the day Wagner's decision was handed down. It was the first time that any federal official found widespread manipulation of the state's energy supplies. El Paso appealed the case, and the FERC commissioners held a hearing in December 2002. It didn't go well for El Paso. By then, Eads's career at El Paso was over. In early January 2003 the company announced that he had resigned. Eads told the *Wall Street Journal* he had "no definitive plans."

A couple weeks after Eads's resignation, Assistant US Attorney John Lewis charged there was a broad conspiracy by El Paso Merchant traders to lie to industry magazines about their trades. By feeding the magazines inaccurate information, El Paso traders could push prices in the direction that benefited their trading book. Three days later, El Paso and its lawyers briefed energy regulators in Washington. An internal investigation had "uncovered evidence that indicated there was systematic price manipulation occurring at El Paso." El Paso said that 99 percent of trades on three of its four trading desks were reported inaccurately. If El Paso's trading position would benefit from higher prices, the traders would report higher prices. If El Paso had a short position that would benefit from lower prices, the traders reported lower prices. The government investigation said it found evidence that traders at several companies routinely reported biased data to industry magazines. Several traders who worked for El Paso Merchant would go to jail, including one who received a fourteen-year prison sentence. In March 2003 El Paso agreed to pay $1.7 billion in cash, stock, and discounted gas to the state and various gas wholesalers in California in the largest settlement ever reached for a regulated company such as El Paso. The next year, El Paso restated its earnings. Most of the revenue and some of the income generated by the El Paso Merchant that reported to Eads were retroactively erased.

Eads was never charged with any legal violations.

When he left the company, he owned 704,000 shares and options worth about $5 million. In 2004, a year after parting with El Paso, he purchased a small Houston financial firm that did a lot of mergers work in the oil patch. His most important client would soon be Chesapeake.

Did El Paso's manipulation of the California gas market end up helping Eads's friend McClendon? High gas prices at the end of 2000 and beginning of 2001 were enormously beneficial to Chesapeake, albeit indirectly. El Paso choked off the main gas pipeline between California and the Henry Hub, a point near New Orleans where sixteen pipelines intersect and national gas prices are set. But Chesapeake didn't produce and sell gas in California, only near the Henry Hub. Would California's market spasms have pushed up gas prices in Louisiana, thereby helping Chesapeake?

I called Vince Kaminski to ask his opinion. A PhD in mathematical economics, he teaches energy markets to MBA students at Rice University in Houston. He was also the former head of research at Enron, where his analytical skills developing energy-pricing models left colleagues in awe. "If you have price impacts in the West, you must have a spillover effect at Henry Hub," he said in a heavy Polish accent. "It can't be avoided. You cannot hermetically separate the markets. Gas can move around. The US market is a system of interconnected vessels." High prices in California would find a way to ripple through these vessels like water seeking an equilibrium. What's more, it was only afterward that it became clear the gas market was being manipulated. At the time, he said, futures traders who arrived at their desks in predawn hours and saw rising prices in the West would integrate this information along with weather patterns and gas storage levels to set prices. If they saw rising prices in Southern California, they would bid up prices elsewhere.

There is no evidence to suggest that McClendon knew what El Paso was doing in California. But El Paso's actions throughout 2000 ended up throwing a lifeline to a struggling Chesapeake all the same. These higher gas prices ended up in hundreds of thousands of home heating bills. The average cost of keeping a home warm with natural gas rose by nearly $400 that winter. More than 1.1 million US households struggled to pay their energy bills, as the combination of rising fuel costs and a rising unemployment took their toll. The rising natural gas prices, which McClendon had predicted, helped put Chesapeake back on track. Its stock price, after bottoming in 1999, began to rise along with interest in the Oklahoma City gas producer.

9

THE FALL OF AUBREY McCLENDON

In late 2002 McClendon delivered a simple message to any audience he could find: the United States was running out of natural gas. "It will be better to be a provider of energy than a consumer of energy for the next twenty years," he told a meeting of energy company executives in Dallas. A couple months later, he pooh-poohed the idea that natural gas production would rise. It was much more likely to fall. The United States produced about twenty-four trillion cubic feet of natural gas a year. "There is no way to get to thirty by 2010," he said. US natural gas prices were headed higher, he told anyone who would listen. Few paid attention. Chesapeake was a small gas producer—it ended the year twenty-first on an annual list of the largest US oil and gas companies—and McClendon was just another slick Oklahoma driller talking up a position that benefited his own bottom line.

Undeterred, McClendon steered Chesapeake on a risky strategic course to take advantage of his vision for gas. He bought other gas producers and leased millions of acres, spending a lot of borrowed money. Then in 2004 he began to notice what was happening in and around Fort Worth with the Barnett Shale.

McClendon assimilated this new information with his existing theory of rising natural gas prices. The United States was short of gas, and these new newfangled shale wells produced a lot of gas. It didn't take much for Chesapeake to begin vacuuming up shale acreage and drilling on it. In 2004 Chesapeake was the eighth largest driller of wells in America. The next year it was the fourth largest. The year after, it was the largest. It would hold on to the top spot for several years. Between 2004 and 2011, it drilled more wells than any other company in the world. Every day, its drill bits began chewing into the earth to start nearly four new wells. The second most active driller wasn't even close.

The small company, named after the East Coast's Chesapeake Bay, where McClendon had spent time boating, became a gas powerhouse by drilling and drilling and then drilling some more. Other companies—XTO Energy, EOG Resources, Devon, and Anadarko Petroleum—followed this path, but Chesapeake set the breakneck pace. Chesapeake's metamorphosis into the world's largest driller required a combination of McClendon's single-minded focus on turning Chesapeake into a large and powerful company, as well as Eads's ability, as the company's go-to investment banker, to find outside financing for his friend's aspirations.

The fly in the ointment was that fracking shale provided more natural gas (and later oil) than McClendon ever envisioned. The industry figured out how to lower costs by drilling wells faster. The wells grew longer, traveling through and opening up more shale, and fracking became more efficient and productive. The cost of producing a cubic foot of gas fell. For a time, profits grew fat and gas extraction leapt upward. This surge of new gas stood in stark contrast to what McClendon and others in the industry thought was going to happen. But McClendon kept his foot on the accelerator, pushing the gas market in the United States from shortage to glut. This change ended up collapsing gas prices. He was hoisted by his own petard.

Sitting on a stage at an industry conference in 2012 with McClendon, Edward Cohen, former chief executive of Atlas Energy, a large Marcellus Shale operator, described the situation succinctly: "We're sort of like the farmers who always want to plant more and more. And the more they plant, the lower prices go."

Aubrey McClendon is a handsome man. He has a prominent chin. He is fit and has an outdoorsy mien. He wears wireless glasses and wears his wavy hair long; it breaks from a side part, flowing in one direction across his forehead, where a few strands turn into large looping curls. The rest undulates back to the base of his skull, where it gathers in a swirled mass that at once looks messy and the work of an expensive stylist. He could be mistaken for retired NFL quarterback Archie Manning.

He is articulate and smart, without being condescending. He can take complex ideas and simplify them. His mind is constantly in motion, digesting information and examining who he is talking to, seeking out ways to connect. Most energy executives are engineers and speak with a tinge of the frustration that comes when the answer is clear to them but eludes others who haven't studied the data. McClendon is a born salesman, leading others to the answer that he has already grasped. His instincts soon become your insights.

This gift of persuasion was on display in early 2012 when he spoke with two visitors to the Oklahoma City campus. Chesapeake, he explained, was a simple manufacturing operation. There were four inputs into this factory. "We have leaseholds," he said, using a term for acreage that the company has leased for drilling. "We have capital," he continued. By this he meant, in a word, money—or, at least, access to money. "We have science, and we have people. Human resources. Those are the only four inputs we have. And we only have one output: oil and gas."

Energy exploration isn't like developing a new cancer drug, where scientific breakthroughs are guarded and patented. The science of shale—and the pioneering techniques used to fracture it— are shared through formal and informal channels. Engineers write papers about characteristics of rocks and publish them for others to see. A fracking site isn't a closed laboratory, sealed behind layers of security and confidentiality agreements. Workers from different companies all participate. The service companies brought in to do specialized work might as well be teenage gossips. The industry employs spies—which it calls scouts—to gather information on its competitors' wells. It is a remarkably open community, a system that has developed over the decades. A wide diffusion of knowledge ultimately helps everyone drill better wells and lower costs.

Chesapeake didn't pioneer either of the two technologies that brought about the shale revolution. Mitchell Energy—and engineers such as Nick Steinsberger—first figured out how to use chemically slickened water to fracture shales. Devon was the first to commit to drilling hundreds of horizontal wells into shales, enabling engineers to frack more of the rock and create giant gas wells. What did Chesapeake do? McClendon figured out how to finance the revolution, tapping into Wall Street's deep reservoir of capital. He sold the revolution to the world's bankers. Of his four inputs, McClendon's signature contribution was attracting capital.

Drilling leases aren't hard to obtain. They require a large check-book. Most landowners are happy to sign for a nice payday. To get the leases requires hiring a lot of landmen, as well as paying their per diems to live out of hotels and drive up and down rural roads. At one point, Chesapeake employed about five thousand contract landmen to lease up areas where it wanted to drill. People and leaseholds are just capital by another name. It takes hard work to chase shale, but capital is the mainspring that turns the gears.

Consider Chesapeake's activities in the Barnett Shale. Eight years after its first shale well, it had accumulated 350,000 acres in northern Texas. It bought some companies to obtain their leases. But the over-whelming majority came in small deals, where its landmen knocked on doors, sat down at kitchen tables, and got the landowner to sign a lease. Chesapeake built its Barnett acreage through small individual transactions. It was a mind-boggling amount of work. It is not un-usual for a large international oil company to negotiate with a for-eign government, and in one deal acquire a half million acres. That's not how leasing works in the United States. Deals must be struck with thousands of individuals. To excel in the shale boom required a small army of landmen and a different kind of energy company.

"It took us two hundred sixty thousand transactions to put to-gether our Barnett asset. How many companies in the world could do one thousand transactions, much less two hundred sixty thou-sand to put it together?" said McClendon. "We have a unique skill set. There is no chance in a million we could go and drill an offshore oil well or an international well as well as our partners could. There is also no chance in a million they could do what we can do onshore in the US."

Building up its Barnett acreage required, first and foremost, people on the ground, meeting with community groups to discuss concerns about where well pads would be located and how many trucks would rumble through a neighborhood. As interest in the Barnett rose, so did prices paid per acre—from $5,000 to $25,000 and above. Even assuming that Chesapeake paid $10,000 per acre, the investment went well into the billions of dollars. And McClendon's ambitions didn't stop at the boundaries of the Barnett. Chesapeake plunged into other emerging shale hot spots: the Fayetteville in Arkansas, the Haynesville in Louisiana, the Marcellus in Appalachia, the Utica in Ohio, and the Eagle Ford and the Permian Basin in Texas. McClendon's pronouncement that it took four inputs can be boiled down to one. It took money, and a lot of it.

His greatest strength ultimately became his greatest weakness. He was so good at raising capital that he created an enormous company, which turned out to be too big to change its direction quickly enough after the financial collapse in 2008. And his ability to raise billions of dollars—not for Chesapeake, but to fund his own personal wells—would eventually lead to his downfall and departure from Chesapeake.

McClendon became convinced by around 2003 that whichever company could grab the most acreage would succeed. Those who dallied would be left behind. By the middle of 2006, he was already declaring victory in the land grab. "The winners for the next ten to twenty years have already been chosen, and the losers will pay the price for years to come. History has shown that Oklahomans did very well in the land run of 1889," he said, a reference to the giveaway to settlers of two million acres. "I believe that history will also note that Chesapeake did very well in the land run of 2000 to 2006." Every lease signed by Chesapeake contained a countdown. Depending on the lease, the company had between three and five years to drill a well. Once the well was drilled and began producing gas or oil, Chesapeake could hold the lease forever—or as long as the well was in operation. If no well was drilled, the lease would expire, and its investment would be lost. Talking to investors in 2006, McClendon boasted that it was meeting its drilling obligations. It had 101 drilling rigs operating, the highest number of any company since 1985, and it was adding rigs.

McClendon wasn't a mere participant in the great shale land grab. He created it. John Pinkerton was the longtime chief executive of Range Resources, the first company to drill wells targeting the Marcellus Shale. "When Aubrey joined the party, all of a sudden the land grab started," he said. This acceleration was unhealthy for all involved. The public wasn't prepared for the onslaught of landmen, regulators were forced to play catch-up, and the industry drilled wells so quickly that common sense was jettisoned at times.

Pinkerton grew concerned about this haste in 2008 after meeting with Pennsylvanians who thought the gas industry was the second coming of coal mining that left the state scarred, with streams poisoned by acid that leached from opened mines. He wanted to be more transparent and hired a team to talk to the public and government officials. He also tried to enlist other CEOs in this effort. He said he couldn't get through to McClendon. "Aubrey was just too schizophrenic. He was everywhere. I couldn't get him on the phone. He totally agreed with me. But he was going so fast, running so many rigs," said Pinkerton. "He was spending all this money and had twenty-five-year old engineers from Oklahoma drilling wells by telephone."

Still, Chesapeake's strategy was copied and envied. Investment professionals wanted to hear how McClendon had pulled the company from the brink of collapse and transformed it into a Wall Street darling. The company's conference calls, where McClendon would talk at length about his view of the energy landscape, were crowded with people dialing in to listen to the Oracle of Oklahoma City.

To finance its rapid expansion, Chesapeake began spending more money than it took in. In its financial results for April through July 2004, the company reported generating $328.8 million from energy operations. It spent $337.9 million on drilling, leasing, and other expenditures. It was, in Wall Street jargon, cash-flow negative. There's nothing wrong with being cash-flow negative—and there are many reasons a company would want to be cash-flow negative. Think of a real estate developer building a skyscraper. It may well run cash-flow negative for a couple years, drawing down a construction loan, line of credit, or cash it has built up. When construction is complete and the building is leased, the developer would become cash-flow positive as rents come in. But the spring of 2004 was only the beginning for Chesapeake. Through the end of 2012, it reported positive cash flow—it made more than it spent—in only three of the subsequent thirty-four quarters. In the first three months of 2012, it reported a $4.1 billion negative cash flow. Over an eight-year span, it spent more than $30 billion on leasing and drilling than it made selling the oil and gas it pumped from its wells.

Building up an inventory of wells to drill turned out to be easy for Chesapeake. Finding capital to finance this operation was tougher. McClendon was endlessly creative in this search. He borrowed money from Wall Street and issued shares; he supersized an existing financial product called the volumetric production payment (VPP), which allowed it to get cash up front in exchange for future gas production. He sold minority stakes in gas fields to foreign companies and he placed complex financial hedges—bets, really—with Wall Street traders on the future movement of natural gas prices.

Facing this seemingly endless demand for capital, McClendon

found an ally in his old Duke friend Ralph Eads. A year after leaving El Paso in 2003, Eads bought a stake in a small Houston firm called Randall & Dewey that specialized on advising companies buying and selling assets in the Gulf of Mexico. But Eads soon steered the company into the more lucrative shale business. McClendon and Eads had remained friends in the two decades since they had graduated from college but had done little business together. That began to change in 2004, when Eads advised Chesapeake on a $425 million acquisition of a small company. Over the next couple years, McClendon talked to Eads regularly about how much capital Chesapeake needed to develop these shales. Intrigued, Eads sat down and scratched out some rough numbers. He figured out how large the Barnett, Marcellus, and a couple other shales were and how many wells would be required. Then he multiplied this figure by how much the wells cost and added in costs to build pipelines. His rough estimate was that Chesapeake and other companies needed $35 billion a year of *external* capital to drill these wells. In other words, $35 billion above and beyond what they generated from selling energy.

"Wow," he recalled thinking. "That's a lot of money." Then another thought occurred to him. There is no way Wall Street will supply even close to that much capital through debt and buying new shares in these companies. Seeing a business opportunity for his firm, which he had sold to New York investment bank Jefferies Group in February 2005, Eads set out to create a new financial ecosystem to find money to drill shale. He helped companies raise capital by selling off older, conventional assets and took a commission on these sales. He found foreign energy companies that wanted to participate in the US shale boom and brought them together with companies such as Chesapeake that were willing to sell minority stakes in exchange for money. Another commission. And Eads went on a global road show, proselytizing to big institutional investors in

China, Korea, India, and Norway on the returns they could achieve from shale. More commissions. Under Eads, Jefferies climbed the league tables—an annual ranking of Wall Street mergers business—like a soccer club with a rich new owner moving up into the Premier League.

Jefferies's pitch material grew into a two-volume, spiral-bound book. It exuded confidence, brimming with upbeat assessments: shale offers "large resources with little geologic risk" and "superior returns." The book's core message was that there were companies with more shale acreage than money to drill. These companies needed partners with deep pockets to help them drill. He aimed the pitch squarely at the heart of the global financial system. Investors around the world have enormous sums of money and seek places to invest that provide high returns. Eads and McClendon were selling an investment opportunity that hit all the right notes. It was a large investment opportunity—$35 billion a year and growing—and it offered strong returns. And the investment was in the United States, which didn't run the risk of governments nationalizing assets or armed bands attacking expatriate engineers.

In a four-year span, from 2008 until 2012, money flowed into North America from Japan, China, Korea, Norway, Australia, India, South Africa, Malaysia, France, and the United Kingdom. Jefferies counted thirty-seven deals that raised a total of $163 billion. By any measure, it's an astounding amount of money, equal to the value of companies such as Coca-Cola or Google. Not all buyers came from overseas. One of the largest deals was when Exxon Mobil bought XTO Energy for $31 billion. Chesapeake alone raised $33.7 billion in a series of deals, for almost all of which Eads served as financial adviser.

People in the oil and gas industry use a metaphor to describe what happened in the mid-2000s. The treadmill, they say, was turned up. What had been a casual pace on the machine—raising money, drilling wells, selling energy—became faster and faster until it was a sprint. While there were many companies and CEOs who sped up the tempo, they were all trying to keep up with McClendon. His desire for "more," as his longtime business partner Tom Ward stated, was the driving force in the rapid development of US shale resources.

Chesapeake discovered a lot of natural gas and leased it up. Its competitors, caught in the company's slipstream, followed its lead. The result was a lot of natural gas—more than anyone had previously thought possible. And when there is too much of a commodity, prices begin to drop. This is especially true when the US economy takes a sharp turn downward. McClendon missed this turn. On August 1, 2008, he said that gas prices would stay in the $9 to $11 range. A year later, they were at $4. "The only thing that went wrong with our strategy was the financial collapse of 2008," McClendon said later. With its heavy debt and drilling commitments, this price decline was a recipe for trouble. It was a major oversight.

"It all goes back to the Austin Chalk," said Dan Pickering, a Houston investment banker whose firm Tudor, Pickering, Holt was involved in the shale boom and its deals. "These guys have historically run until they fall off the edge of the cliff. In the Austin Chalk, the geology wasn't there. This time around, it was financial leverage. You want to push and push right up to the very edge and not go over. Figuring out where the edge is, that's hard."

Chesapeake had another large source of revenue that helped the company grow. McClendon placed aggressive bets on future movements of natural gas prices. He was a good trader, generating large profits. Between 2001 and 2011, Chesapeake reported $36.7 billion in revenue from selling the oil and gas it produced. Over the same period, McClendon generated $8 billion in trading profits. For every dollar that Chesapeake took in over that period, eighteen cents came from Wall Street trading, not wells. Over this span, McClendon had seven years of gains from trading and only four down years. And his good years were very good, while his down years didn't generate particularly large losses. Only once, in 2004, did he generate a loss of more than $100 million. Five times, his trading resulted in billion-dollar gains. This extra cash was critical to Chesapeake's rapid growth. Chesapeake used it for aggressive leasing, speeding up the pace of shale development and forcing other companies to ramp up their spending if they didn't want to be left empty-handed.

McClendon also traded on his personal account and for a hedge fund he set up with Tom Ward in about 2004. I stopped by the offices of the hedge fund, called Heritage Management Company, one day in 2007. It was in a 1930s skyscraper on Fifth Avenue in Manhattan. There was no receptionist, so I walked through the foyer into a small room where I startled four traders seated around a large table, staring intently at their twin computer monitors. They didn't want to talk about the company and ushered me out of the office. When I later asked McClendon about his outside investments, he replied, "I just play the stock market, I play the commodity markets. Sometimes I win, sometimes I lose." He said later that the fund helped give him insights into the market, which helped him manage Chesapeake better.

Still, the arrangement was unusual. McClendon was the CEO of an expanding natural gas company that grew to control a substantial portion of the domestic gas market and was active in futures trading. McClendon was also actively trading the same energy contracts for his personal account. Were Heritage traders allowed to place bets on future movements of commodity prices *before* Chesapeake secured its price hedges or publicly announced increases or decreases in drilling? The potential existed for "front running," when a trader buys or sells commodities before executing his company's trades. This could allow McClendon to profit from advanced knowledge of price movements. There is no evidence that anyone oversaw the operation, at least anyone besides McClendon.

By 2008, McClendon and Chesapeake were whales in the futures market, a massive electronic exchange where contracts for billions of barrels of oil and trillions of cubic feet of gas are traded. How big is difficult to say with precision. Trading positions on the New York Mercantile Exchange are zealously kept out of sight by a club of investment bankers, hedge fund managers, physical traders, and speculators. Those in the club, where membership requires access to billions of dollars of collateral, share snippets of gossip about who is betting prices will rise or fall. But for those outside, information can be tough to come by. This veil of secrecy has fallen only one time, and the snapshot of the investors and traders setting energy prices in the summer of 2008 showed that McClendon and several close associates were among the largest participants.

US Senator Bernie Sanders, an independent from Vermont, pulled the curtain back when he released a confidential list of futures market speculators compiled by the US Commodity Futures Trading Commission, a disclosure that sent futures traders into conniptions. The list of who held the natural gas contracts was a who's who of global capitalism and energy. In descending order, they were BP, Barclays, Morgan Stanley, JPMorgan, and Shell. The sixth largest was the reclusive Tulsa billionaire George Kaiser, who owns a private gas exploration company, and who lent money to McClendon and often talked to him about market fundamentals. The eighth largest was John Arnold's Centaurus Advisors, a major Houston hedge fund in which McClendon had invested.

Chesapeake Energy held gas contracts to make it the seventeenth-largest participant in the futures markets, by far the largest position for an energy company of its size. Only a few individuals made the list. After George Kaiser, there was T. Boone Pickens (number 12). The next individual on the list was Aubrey McClendon, followed by his longtime business partner Tom Ward (numbers 52 and 53). Taken together, this cluster of Texans and Oklahomans all connected through McClendon and including Ward, Chesapeake Energy, Centaurus, and Kaiser held about 10 percent of the total natural gas futures contracts, a large accumulation on an exchange that helped set the continental price of gas. The largest single trader—BP—held only 7 percent.

The second week of October 2008 was abysmal for anyone invested in the global stock market. The Dow Jones Industrial Average fell 18 percent, it worst five-day span since the depths of the Great Depression. On Friday, the tenth of October, the Dow—a group of the market's largest, most important companies—traveled a vertiginous path in its most volatile day ever. Even in this market, Aubrey McClendon had a particularly bad week.

Shortly before noon on that Friday, Chesapeake disclosed that McClendon had been forced to sell off virtually all of his shares in the company. He had borrowed substantial amounts of money from brokerage firms, using the value of his 33.5 million shares in the company as collateral. When the value of these shares fell, triggers embedded in these loans were crossed, forcing McClendon to pay back his lenders. McClendon had maintained a margin account for years, using his Chesapeake shares as collateral to trade stocks, bonds, and commodities. He had entered the week with a 5 percent stake in the company worth $2.32 billion a couple of months earlier.

His stake in Chesapeake had always been a point of pride. He linked his personal wealth to shareholders because he was a shareholder. After the sales, he owned $32 million in Chesapeake shares. "I got caught up in a wildfire that was bigger than I was," McClendon said over that weekend.

His financial house in tatters, McClendon borrowed $30 million a week later from Centaurus Capital, the Houston hedge fund that traded in natural gas and other commodities. Centaurus was run by John Arnold, a thirty-six-year-old former Enron whiz kid trader and billionaire hedge fund manager. As collateral for the loan, McClendon put up his stake in a Centaurus fund. The filing, in the Oklahoma County Courthouse, went unnoticed for years. In retrospect, it is a stunning document. Aubrey McClendon, the CEO of one of the largest natural gas producers in the United States, was investing in a hedge fund that amassed enormous profits trading natural gas. The fund owned considerable gas assets.

This debt was the first, but not the last, that McClendon accumulated beginning in late 2008. He borrowed from individuals and firms with large energy-trading operations. He pledged his ownership stake in the Oklahoma City Thunder to Bank of America. To George Kaiser, he pledged his venture capital company, then a half interest in some of his privately held oil and gas wells, and later his collection of petroleum memorabilia. To Goldman Sachs, he pledged a large portion of his wine collection, which ran seventy-eight pages and included two six-liter bottles from the famed La Tâche vineyard in Burgundy and more than two hundred bottles of cabernet sauvignon bottled in 1983 by Château Margaux.

That "more" was McClendon's governing philosophy is clear from even a cursory look at both his and the company's assets. Chesapeake had acquired leases to drill on more of the United States than any other company had ever amassed—15.3 million acres—and bought a large three-story office building in Oklahoma City just to house all of the land records. Along the way, Chesapeake started buying up property in northern Oklahoma City for a corporate headquarters. Starting with a single building, Chesapeake snapped up 111 acres and built a dozen three-story redbrick buildings, with gray roofs and white dormer windows. The campus fitness center was state of the art, with squash courts and a basketball court good enough for the New Orleans Hornets of the NBA to use when they evacuated following Hurricane Katrina in 2005. Chesapeake also built a shopping center across the street, attracting the first Whole Foods Market in the state.

McClendon had also personally gone on a buying and building spree—much of which became collateral for his debts. He acquired an ownership stake in a television station and several restaurants in Oklahoma City. He created a large tree farm and opened a roadside restaurant on Route 66 with a sixty-six-foot-tall soda bottle that lights up at night called Pops that sells five hundred different varieties of soda. Near where his wife grew up in Michigan, he planned a massive lakeside development and ended up in a lengthy battle with the local township over zoning. Near Gull Lake, Minnesota, he housed his collection of eighteen antique wooden-hull motor boats, including a 1936 Ditchburn yacht valued at $1.5 million. He acquired a vacation home in Bermuda and planned another in Hawaii.

His largest personal collection, however, was the stake in wells he built through the Founders Well Participation Program, the perk that allowed him to purchase a 2.5 percent stake in nearly every well that Chesapeake drilled. In 2008 it drilled 1,733 wells. McClendon elected to participate—and the bill for his participation kept growing. He was likely already the largest single owner of oil and gas properties in the United States. Four years later, Chesapeake disclosed that his small slices of thousands of wells produced the equivalent of the 147 million cubic feet of natural gas a day, enough to supply all residential customers in Connecticut or Kentucky.

For many executives, the fall of 2008 was a humbling time, a time to reassess business strategies and survive the financial storm. Chesapeake charged forward, but investors were increasingly unhappy. In late November, just a few hours before the market's Thanksgiving holiday, Chesapeake filed forms with the federal security regulators to allow it to issue $1 billion in new shares. Share prices plunged when trading reopened. Within a few days, Chesapeake reversed itself and said it would slash spending. "These filings were a mistake," McClendon said in a conference call with investors. "I apologize for that and ask your forgiveness for it."

Chesapeake's shares fell by 59 percent in 2008, and McClendon exited the year with a dented ability to convince investors that he could guide the company in the right direction. But the Chesapeake board members decided to reward McClendon. In early 2009 they gave him a new five-year contract and a onetime $75 million "retention" bonus. The money would help him pay his obligations to the Founders Well Participation Program. The company also agreed to buy McClendon's collection of antique maps for $12.1 million. The board's generosity made McClendon the highest-paid CEO in the United States in 2008.

The maps, the special retention bonus, and other problems came to a head in June, when the pro- and anti-McClendon camps gathered for Chesapeake's annual shareholders' meeting. These events are usually highly scripted affairs, a cavalcade of praise for the executives. That year's meeting was tempestuous. One longtime investor assailed McClendon for confusing rising gas prices with his own success. "Your greed and your ego took over, and you bet the farm that your success would continue," said Jan Fersing, a Fort Worth businessman. "So, your two-billion-dollar fortune was not enough; you wanted more. But this time your hand got stuck in the cookie jar, and you couldn't let go until your own cookies were taken in the process. And after your embarrassing losses, but with a carefully picked and extremely well-compensated board of directors, Chesapeake shareholder funds were partially used to cover your losses."

McClendon remained calm.

"I've worked one hundred hours a week, at least, since 1989 building this company. I've sacrificed a lot to do that. I sacrificed five hundred million dollars that I lost last fall as a result, not because of bad decisions but because of things beyond my control in this country's economy and with regards specifically to natural gas prices. So, I'm sorry that you find me as egocentric and greedy. But, I'll tell you there's not a harder-working guy out there who thinks every day about how to create shareholder value," he replied.

After a couple more shareholders praised Chesapeake, the final member of the audience spoke. "Hi. My name is Ralph Eads. I'm a longtime shareholder of the company. Virtually everybody in my family owns the stock. It's been, for like some of the other folks here, transformative for me. I am a longtime professional in the oil and gas business; I know a lot about the industry, and I know a lot about this company. And I'll just say it is the finest company, and Aubrey is the finest CEO. And I'm here today—I made the trip from Houston—to thank the board for the compensation arrangement they did with Aubrey. It kept him—it will keep him in the saddle for the next five years, and that will make everybody in the room and all the shareholders a lot of money. So, I know it was a difficult decision for you guys, and controversial, but I want to tell you I appreciate it a lot. Thank you."

McClendon thanked him for his comments, without noting his longtime friendship or business relationship, and moved on.

Chesapeake continued its strategy of spending more than it brought in—and closing its funding gap by selling off assets. Over the next couple years, it sold six volumetric production payment deals for $3.6 billion. These agreements with large Wall Street banks provided Chesapeake immediate cash in exchange for future gas production from its wells. Chesapeake used the cash it received to fund a still-ambitious growth campaign. As Chesapeake did more VPPs, the amount of gas it promised to deliver grew—as did its obligation to pay to pump this gas and deliver it without getting any money. Few paid attention to these deals. One exception was Pawel Rajszel, a young security analyst in Toronto, who issued a scathing report to his clients in April 2010, arguing that these deals put the company in significantly more debt than it cared to acknowledge. Pointing out that Enron had pioneered the VPPs a decade earlier, he wrote, "Due to what we consider an accounting loophole, Chesapeake is effectively able to hide its VPP liabilities from its balance sheet—something even Enron Oil and Gas Company did not do." While investment professionals didn't pay much attention to Rajszel's prescient warning, Chesapeake's debt-rating agencies agreed these deals should be accounted for as debt. McClendon spent considerable time trying to change their minds. When he didn't, he told investors these debt watchdogs were wrong. "They see VPPs largely as debt, which is kind of nutty," he said in 2010.

Chesapeake also sold off assets and, with the help of Ralph Eads, lured foreign investors as partners in Colorado, Pennsylvania, Louisiana, Arkansas, and Texas. Beginning with a deal to sell 20 percent of its acreage in the Haynesville Shale, in northern Louisiana, to Plains Exploration & Production Company for $3.16 billion, McClendon attracted buyers for the land he had built up. By the beginning of 2012, he had struck deals to sell off minority positions in Chesapeake assets worth more than $20 billion. Other companies copied this strategy of bringing in joint-venture partners, but Chesapeake pioneered the practice and was the most active. Chesapeake was good at using its landmen to snap up acreage in newly discovered shale fields before anyone else and then selling off a chunk to companies from China, France, or Norway to recoup its investment—sometimes even before it drilled more than a handful of wells. By 2011, the shale revolution had succeeded at finding too much natural gas. Prices were falling. After promising investors it would live within its means, McClendon said that Chesapeake needed to capture new opportunities. And it kept spending more than it made, leasing and borrowing and selling what it could.

In August 2011 Aubrey McClendon appeared on *Mad Money*, a cable financial news show on CNBC. He was there to boast about a newly discovered shale—the Utica—in Ohio. It was a typical performance. The Utica is "one of our biggest discoveries in US history." Maybe twenty-five billion barrels of oil and petroleum liquids. Really big. Chesapeake had drilled in total only fifteen wells in the Utica—and only six modern horizontal wells. From this small sample, it extrapolated that the discovery could be worth between $15 billion and $20 billion. It was a bold statement, considering the entire company was worth about $22.6 billion at the time.

Not long after, Chesapeake began to turn the acreage it had leased into money. Before the end of the year, two deals came together. One was a joint venture with French oil major Total to sell a quarter of its Utica acreage in ten counties in eastern Ohio. Chesapeake had also created a new subsidiary, placed most of its Utica acreage into it, and then sold $500 million in shares of this subsidiary to EIG Global Energy Partners. EIG, which ultimately spent another $750 million on CHK Utica shares, received a nice 7 percent annual dividend on its investment, plus the first 3 percent of revenue from the first 1,500 wells drilled in the Utica. Eads represented Chesapeake, as financial adviser on the joint venture and as adviser helping to connect Chesapeake with EIG. It was the twentieth and twenty-first transactions that Eads had undertaken for Chesapeake, cumulatively generating $28 billion for the company, since the beginning of 2007.

When Chesapeake released details of the EIG deal, it spelled out a mind-numbingly complex financial transaction. What it didn't disclose was that a couple weeks after McClendon sold off a chunk of Chesapeake's hottest new acquisition to EIG, he was arranging a personal $500 million loan from EIG. (The Reuters news agency disclosed this in an April 2012 story that walloped Chesapeake's stock price.) The loan helped McClendon continue buying his 2.5 percent share in Chesapeake wells. And he was using his ownership in these wells as collateral. Henry Hood, Chesapeake's general counsel and another Duke classmate of McClendon's, told investors that everything was aboveboard. "The board of directors is fully aware of the existence of Mr. McClendon's financing transactions," he said.

This assurance did nothing to squelch investor discontent. The loans turned out to have been substantial: up to $1.4 billion to cover several years of participation in the perk. In 2011 McClendon had to come up with about $450 million to participate in the all-or-nothing program. The federal government opened two separate inquiries into whether the company had disclosed enough information— and whether McClendon had reported his income appropriately.

Chesapeake took the unusual step of recanting Hood's statement that the board was in the loop, eight days after it was issued. As it turned out, the board didn't know about the loans. It "did not review, approve or have knowledge of the specific transactions," according to a press release. By the end of the year, Chesapeake demoted Hood. The board, pushed into exerting its influence, negotiated an early end to the Founders Well Participation Program. Amid calls for better oversight, Chesapeake announced on May 1, 2012 that McClendon would step down as chairman but kept his chief executive role.

Large shareholders, however, were not placated, and McClendon appeared to dislike his diminished role. In late January 2013 Chesapeake's new board of directors issued a two-page statement. McClendon was stepping down as CEO and leaving the company he had cofounded and steered for more than two decades. A couple days later, it clarified that he had been terminated. Former Conoco chief executive Archie Dunham, who had taken over as chairman the previous June, assured employees in an email that Chesapeake and the unique culture that McClendon created wouldn't go away. The fitness center would remain. And the gourmet company cafeterias would not be sold to a food-services company. Within a few months, however, Dunham was steering Chesapeake in a new direction that was significantly less aggressive than it had been in the past. It was nothing less than a repudiation of the strategy McClendon had pursued.

McClendon left Chesapeake on April 1, 2013. Less than a month later, he sent an email to industry contacts. He was starting a new company called American Energy Partners. "I am interested in being contacted regarding onshore US assets," he wrote. "In particular, I will be looking for deals with a lot of drilling left on them and will also consider undeveloped acreage deals—plus, I am not scared of natural gas."

In 2003 McClendon had said there was "no way" to build an economy that consumed thirty trillion cubic feet of natural gas by 2010. But a couple years later, the United States was producing thirty trillion cubic feet a year. Companies vented some of it into the atmosphere and used some of it to power the machinery of fracking. It was still too much. America was consuming only twenty-five trillion cubic feet.

McClendon remains an unrepentant gas enthusiast and optimist. Asked if he had any regrets about his tenure at Chesapeake and whether he would do anything differently, he replied: "Nothing, I loved every minute of it." After 2008, prices remained low enough, he contended, to stimulate more people and companies to commit to using natural gas. Now that demand is catching up with supply, it is again a good time to drill wells and produce natural gas. With a typical mix of boldness and far-reaching rhetoric, he said: "The future will once again belong to producers over the next twenty to fifty years."

10

CELESTIA

Fracking is changing the world, but some places are shoulder-ing more of this change than others. To understand better what Aubrey McClendon and Chesapeake have created, I returned to the Farm for a week to spend time traveling around Sullivan County. Chesapeake had leased thousands of acres and drilled nearly one hundred wells, including one on my parents' property.

The debate over fracking here isn't abstract. US energy secu-rity and climate change don't come up that often in conversation. When residents of Sullivan County talk about fracking, they talk about their water and land as well as the trucks on the roads. What I learned—and would have realized long ago if I had thought about it—was that my parents' approach to the land was out of step with their neighbors'. They planned to keep the land untouched and wild. Many of their neighbors worked the land for their livelihood.

Over the years, my parents had inched toward the local way of thinking. They allowed a local timber company to harvest some of the trees. They posted No Hunting signs on the property, but within a few years relented and let neighbors come in during white-tailed deer season.

There are people opposed to fracking in Sullivan, but they seem to be recent arrivals. As I spent time talking to a variety of residents, I came to believe that the longer a family has lived in the county, the more it was willing to allow gas drilling. As I talked to locals, I began to realize that the county was gripped in a form of seller's remorse. They had agreed to lease, but now had second thoughts about their decision. The sheer magnitude of the change was what worried most people. Nearly everyone had signed leases and turned over the future of the county to outsiders.

Not long after I arrived, I sat on the porch of a house near the Farm, drinking lemonade with a couple who had lived there for decades. I had bumped into the husband, Wylie Norton, at the courthouse that day. When he found out my parents' cabin was a mile down the road, he invited me over. Wylie is a county commissioner, and his wife, Melanie, worked at the county's small historical museum. As a cat tried to leap into my lap, Melanie relayed a piece of the county's history that I had never heard, but it had odd parallels to my parents' story and the gas boom unfolding around us. "Have you ever heard of Celestia?" she asked, between drags on a cigarette. When I replied I had not, she told me to stop by the two-room museum.

Many parts of the United States boast they are God's country, but only Sullivan County has the legal paperwork to back up the claim. The roots of a most unusual real estate transaction began in 1850, when a husband and wife from Philadelphia named Peter and

Hannah Armstrong bought land in the middle of the county. Like my parents a century later, the Armstrongs were lured by Sullivan County's cheap real estate and the promise of escape from the hustle and bustle of the city. The similarities end there.

Peter Armstrong was a divine communist and a messianist who believed that Jesus Christ would soon return to earth. Following the book of Isaiah, which says the "Lord's house shall be established in the top of the mountains," the Armstrongs bought several hundred acres atop one of the county's mountains and began to build the city of Celestia, a community that would be pure of sins and follow divine, not human, law. They planned a city of thousands centered around a vast four-story temple with a pyramid atop it that reached toward heaven. This grand vision was never realized, but a handful of Philadelphians followed them and built homes.

Celestia might have disappeared—a footnote in the history of nineteenth-century religious revivalism—if not for a curious incident that unfolded in 1864. During the Civil War, one of Armstrong's followers sought a religious exemption from conscription into the Union army. Armstrong wrote President Abraham Lincoln a letter and received a speedy reply, approving the request. Building on this success, he decided to seek exemption from property taxes also. The matter was referred to a local board, which devised a clever response. If Armstrong claimed that his faith led him to seek separation from earthly affairs in the wilderness, then surely holding a deed to the property was inconsistent with his faith. The board gave Armstrong a choice. He could either give up his faith and keep his property or relinquish his property and keep his faith.

Armstrong, however, was more than up to the challenge. His response is recorded in the Sullivan County Courthouse, in a deed dated June 14, 1864. He transferred Celestia, about six hundred acres of land and, incidentally, the attached mineral rights as well, to God.

"We do, by these presents, deed, grant and convey to Almighty God, who inhabiteth Eternity, and to His heirs in Jesus Messiah, to the intent that it shall be subject to bargain and sale by man's cupidity no more forever, all our rights and title (by human law) and claim of any nature soever in or to, of that certain tract of land," the one-page document notes. He likely drew inspiration from another portion of the Old Testament that states, "The land shall not be sold for ever: for the land is mine."

It was an elegant solution. The six hundred acres were no longer owned by people of this earth. Celestia remained put. But there was one problem; God was a tax delinquent. Armstrong and his followers may have been trying to build in the wilderness a new city that glorified God, but local officials held a different view of real estate. They believed that Sullivan County was there to be worked. Small subsistence farmers were settling all over the county, clearing out the hemlock forest to harvest the bark, which was used in nearby tanneries. Soon railroads would enter the county, to transport its coal north and blocks of ice south. The United States was in the midst of an energy transition; coal would soon displace wood as America's primary fuel and fire the nation's industrialization. Pennsylvania embraced its role as a resource provider and became a leading producer of coal, iron, steel, timber, and petroleum. Sullivan County required tax revenue to build roads and supply services to its population, which increased by nearly half between 1860 and 1880.

By 1876, the Sullivan County treasurer had begun to demand Celestia's back taxes, but Armstrong refused to pay. There are two stories of how the showdown ended. One is fanciful and the other probably true. The first is that the sheriff went to Celestia to take its sheep as payment for taxes owed. Armstrong gathered his followers to witness a miracle. The sheep, he said, wouldn't follow the sheriff. But they did follow him, and soon most of Armstrong's human flock departed as well. The other version of events is that the treasurer sold 350 acres at a sheriff's land sale. The buyer was Armstrong's son, who paid $33.72, an amount that covered the tax bill.

Whichever version is correct, within a few years, Celestia was in terminal decline. The Armstrongs' bid to sever the land's earthly ties from commerce and government had failed. Their son sold the land to a lumber company, and the forest was cut down for its bark. The Armstrongs couldn't break free from nineteenth-century trade and taxation. My parents also found they couldn't escape from twenty-first-century energy exploration, and neither could their neighbors. Even the holdouts are now surrounded by pads, rigs, and pipelines. The rumble of trucks is too ubiquitous to escape, the lure of jobs servicing this new industry too great to ignore.

The energy industry juggernaut has rolled through Sullivan County. The only real question left is what the county will look like after the gas industry has come and gone. Sullivan County was nearly denuded in the nineteenth century as its old-growth hemlocks were felled, but the forest returned. Today a lush canopy of second-growth pine, ash, and oak trees covers the mountains. The coal mine is, for the most part, gone too. A strip mine was filled in and reclaimed. Both the tanneries and the coal mine dumped toxins into streams, but most of the rivers run with clear, drinkable water. Still, the arrival of the gas industry has left some locals wondering whether the land and water will recover this time. "Our forests were not harvested correctly and our coal was not harvested correctly; left us with acid runoff. We can't do that three times in a row," said Erick Coolidge, a dairy farmer and supporter of gas drilling. "Where is our plan? If we don't do this right, what the hell have we done?"

———

Sullivan County is the second smallest of Pennsylvania's counties, with 6,100 residents. It is set amid the Endless Mountains, part of the Appalachian Mountains chain. Deer outnumber year-round residents, and black bear sightings are common. At the beginning of the twentieth century, the town of Eagles Mere was a country resort for New York's elite. The posh hotels are gone, replaced by small inns and bed-and-breakfasts. There is no McDonald's or Wal-Mart, and you can't get cell phone reception in large swaths of the county. There are a dwindling number of dairy farms but many remaining fields of alfalfa, corn, and hay. And there are trees—millions of hardwood cherry, maple, ash, and oak trees—that are chainsawed, cut into boards, and exported around the world to be turned into flooring, cabinets, and caskets. The local timber industry prides itself on sustainable forestry, not clear-cutting.

Below it all is the Marcellus Shale. Generations of farmers, tourists, religious seekers, and city dwellers were drawn to Sullivan's soil, wilderness, and beauty. What was below the forest floors attracted the petroleum industry geologists beginning around 2006.

The first wave of landmen targeted the largest landowners. Betty and Milo Reibson signed a gas lease in 2006. They are one of the remaining dairy farming families. Their herd of one hundred cows is one of fourteen in the county, down from nearly three hundred at the peak. In their sixties, they remain active farmers even as they turn over more day-to-day responsibility to their son, the third generation of Reibsons to operate the farm. Their pretty two-story white house sits a few feet off the road, up the hill from the barn and two weathered gray silos. On an overcast morning, most of the cows remained in the barn. Eight heifers and cows, the pregnant members of the herd, grazed on a hill behind the house, hemmed in by an electric fence.

The landman's first offer was $25 an acre to drill on the Reibsons' land. The conversation didn't go well. He came and sat at their wooden dining room table with a topographic map. When they asked to see it, he refused. Milo offered to walk the young man around the property to give him a sense of the layout of the fields. The landman, who worked for Anadarko Petroleum, said this wasn't necessary. This refusal put the Reibsons on edge. They weren't against gas drilling, but business wasn't done this way in Sullivan County. The land, its contours and soil and water sources, was all that mattered. If they planned a major change on their property, they would drive down the road to their neighbors for a discussion. The out-of-state landman was operating from a different playbook.

When I arrived at the Reibsons' house on an overcast day in July 2012, Milo greeted me in a ratty and torn blue checkered shirt. Betty descended the staircase a couple minutes later, asking my forgiveness for her lateness. Her hair was perfectly in place. She wore pressed black slacks and a white blouse. She has a direct manner and a way of looking at people that holds their attention, a political skill that served her well as she rose from being a township supervisor to county commissioner. "If I was to train people to go out and talk to people about their biggest asset and most heartfelt asset, I would train them with more sensitivity," said Betty.

Despite the culture clash with the landman, the Reibsons didn't dismiss the offer outright. Farmers had signed leases on and off for decades in Sullivan County. The terms were small, a couple dollars an acre, and no drilling ever took place. Farmers grew accustomed to what was basically a hassle-free stipend. The Reibsons soon realized that this time was different. The money being discussed was several times higher than in the past. As negotiations dragged on for six months, Milo talked the landman up to $85 an acre. (The lease was for 245 acres, so the sign-on bonus was more than $20,000.)

With this kind of money on the table, they expected there would be a well drilled. Milo and Betty negotiated an addendum to the lease that gave them the right to consent to the location of any well on their property but required they not withhold permission "unreasonably." The Reibsons signed a lease with Anadarko. Two years later, Anadarko sold 50 percent of its interest to Chesapeake and designated the Oklahoma City company as the operator.

Chesapeake acquired the majority of the acreage across the northern tier of Sullivan County. The land is divided into thousands of plots, few of which resemble a shape found in grade school geometry. The property lines tend to follow what can be cultivated. Fields have been sculpted into the undulating hills, bought and sold by farmers, leaving behind a crazy quilt of real estate records. Rather than drilling on every property, Chesapeake pooled the plots into units. Each unit was about 640 acres, roughly one-third mile wide and one and a half miles long. The units are long rectangles, running on a southeast-to-northwest axis. As the units were drawn up, property owners received a share of each unit commensurate with the number of acres they owned. In early 2011 Chesapeake created the Phillips Unit. The Reibsons owned about 61 acres in the unit, entitling them to royalties on 9.5 percent of the gas in the unit. The company created an adjacent unit that contained 190 of the Reibsons' acres. Following convention, Chesapeake named it after a landowner and called it the Milo Unit.

The way the Milo Unit was drawn up, the Reibsons' parcel sits at the dead center. That's where Chesapeake wanted to locate its pad: a flattened, compacted piece of earth where it could assemble the machinery needed to drill a cluster of six wells. These wells would then spread out like a spider with six long legs, with the pad as the cephalothorax. Three wells would head to the northwest, running parallel to one another; the other three to the southeast. Once each well had been fracked, the cracks would spread out from each well, covering the units and allowing it be drained of gas in a systematic, efficient manner. Little, if any, of the buried Marcellus Shale would be untouched. The layout was scientific, systematic, and thorough. It was a factory approach to exploiting the shale, the machines placed in the exact spot to maximize production and minimize costs. And it had little regard for the surface. According to Chesapeake's logic, the pad should be located in the center of the unit. If it were placed at the top of the rectangle, three of the wells would need to be extra long and the other three short, raising costs of the entire operation.

In April 2010 Chesapeake mailed the Reibsons a surface-use agreement proposing to place the Milo Unit pad on a hill behind their house. An enclosed sundeck on the back of their house contains their television, couches, and a mounted black bear that Milo shot. The sundeck looks out across a picnic pavilion, a small patch of blueberries, and up a small hill. The left flank of the hill would be the new drill pad. The proposed access road would go right through their yard, passing within a couple dozen feet of their bedroom. The Reibsons said no. In addition to the trucks rumbling past their house, the pad sat atop their water. A spring flowed out of the bottom of the hill, where they collected the water in a reservoir and pumped it back uphill to the barn and the house.

The Reibsons were furious. Chesapeake's proposed well location required drilling through their water supply. Any accidental spill could contaminate it. Milo had negotiated the addendum so this wouldn't happen. Milo Reibson is a lean man with the body of someone who has engaged in physical labor his entire life. He is tall and bald, with a trim mustache. He looks like an ex-ballplayer who has stayed in shape. Sitting in his living room, fiddling with an unopened Mountain Dew, he explained that he'd insisted on the addendum specifically "so I would have some say-so, and they wouldn't sit on my water or my very best fields." The Reibsons' anxiety about their water is understandable. The concern was repeated all across the county—and everywhere in rural Pennsylvania where drilling into shale has taken place. Without a good source of groundwater, a dairy farm won't survive. The spring is the only source for the four thousand to five thousand gallons that the Reibsons' herd of milk cows drinks daily. It also provides water for the couple's kitchen and showers. There are no municipal water connections nearby. If the spring loses its freshwater, an auctioneer to sell off the herd would be close behind, ending the Reibsons' six-decade run of dairy farming.

The Reibsons rejected the surface-use agreement. They didn't think they were being unreasonable for a minute. And so began a long battle between the Reibsons and Chesapeake over siting the well pad that would end up in federal court in Scranton, Pennsylvania. Lawyers for Chesapeake, the plaintiff, argued that the Reibsons were being "arbitrary and unreasonable" and wanted to stop the company from drilling on their property. The Reibsons' lawyer countered that it was Chesapeake being unreasonable. The company concocted its Sullivan County map of well locations and unit boundaries "to maximize its production and profits—irrespective of whether those sites caused any harm to the landowners' business or quality of life."

This was a clash between two different modes of capitalism. Chesapeake was developing a factory in rural Pennsylvania for the extraction of gas molecules. Its maps and leases were all sensible and legal, designed to lower costs and increase profits. From faraway Oklahoma City, the rectangular blocks being created for Sullivan County made all the sense in the world. They were the output of geologists who studied the alignment of the shale, the angle of natural fractures, and the optimal placement for man-made fractures. But from a farmer's perspective, Chesapeake was both blind and arrogant. Placing a well pad atop the farm's only reliable water source? Plotting it out without consulting with the landowner? It was downright disrespectful and out of step with a modern dairy farm operation.

Over the next few months, the Reibsons proposed alternative locations. And so did Chesapeake. The exact details of who rejected which location was disputed in the lawsuit. Milo Reibson said the gas companies haven't bothered to understand what's important to farmers. His farm has 150 acres of heavy clay soil and 75 acres of red shale-flecked soil. The rest is forest. The red shale is the good stuff, generating higher yields of corn and alfalfa. He kept trying to steer Chesapeake's pad onto his clay soil. If he was going to lose a handful of acres, taken out of production forever, it only made sense to lose the bad soil. Chesapeake kept coming back to the good soil.

The conflict was soil versus rock, farming versus petroleum engineering. There were two conflicting ways of viewing the property. The Reibsons looked at their farm and saw the cows, land, and water and how they all fit into a long-term management plan. Chesapeake saw the gas potential and desired to drill wells and accelerate the depletion of the resource to speed up its payout.

I asked Milo if I could see where the pad would end up being built and where Chesapeake had proposed putting it. Milo Reibson pulled on a pair of work boots. We walked up the hill behind the house. He warned me to keep my distance from the pregnant cows, who had stopped chewing and were watching us warily. When we reached the crest of the hill, below us was a pond with a couple ducks and the spring that had provided an uninterrupted supply of freshwater since the 1960s. The cows turned this water into millions of gallons of milk sold in grocery stores. He told me the story of coming home one day to find stakes in the ground. A gas company surveyor had trudged up the hill and laid out where the pad would be. Milo chuckled. There was an ornery bull along with the pregnant cows that day. It was a wonder the surveyor didn't get chased, he said.

We headed toward a barbed wire fence that ran along a line of trees. On the other side of the fence was one of his fields—a good one with red shale. It's farther from the house than the first location Chesapeake wanted to use, but where it had agreed to place the pad. He expected to lose the field forever. There's the Marcellus, he explained, but even deeper is the Utica Shale. In Ohio, the Utica Shale has just begun to yield oil wells. After the Marcellus wells have been drilled and fracked, Chesapeake or some other company that Chesapeake has sold its lease to will return and explore the Utica potential. "They will never leave this pad," he said. "Once they get you locked up, they never leave."

Betty Reibson had told me that they couldn't afford a lawyer—the federal government had been cutting the regulated price at which that milk can be sold—but hired one anyway. She told me she hoped that they could all find a way to coexist. "We're going to have to," she said. "They're here, and we're here."

A few weeks after my visit, months of negotiations with a federal magistrate serving as the middleman between Chesapeake and the Reibsons produced a tentative agreement about the pad's location. The lawsuit was settled out of court in early November. A week later, Chesapeake applied to the state for permission to begin drilling a well.

Drilling rigs were slower to come to Sullivan County than to neighboring counties. To the north, Bradford County had 1,821 shale wells, drilled or permitted, as of the summer of 2012, and Tioga

County had 1,138. To the west, Lycoming County had 926. Sullivan County had 166. The reason Sullivan has seen less activity is that there are no pipelines to get natural gas out of the area. By the end of 2012, the Marc 1 Hub pipeline was assembled, welded together, and buried at least three feet deep. A thirty-inch-high pressure gas pipeline, it runs for thirty-nine miles, crossing more than one hundred streams and creeks, as well as the Susquehanna River. It is a new major artery in the nation's gas grid, connecting two giant pipelines that carry gas from the Gulf Coast up into the populated Eastern Seaboard. As it crosses through Sullivan County, it takes feeds from new wells that have been drilled but during the summer had yet to be turned on. The Marc 1 began operations on December 1, 2012, and in the days that followed, gas began to flow for the first time from the county. Some of it heads north to connect to the Tennessee or Millennium pipelines and from there to New England. The gas can also head south into a spur of the Transco pipeline, the largest in the United States by volume, which feeds into the New York metropolitan region. Plans are afoot to add another pipeline leg that will allow gas from Sullivan County to head to consumers from Washington, DC, to Boston.

As the large Marc 1 inched closer to completion, gas companies scrambled to put in a grid of smaller pipes that will connect each well to eastern consumers. During my stay, these connector pipes were spreading across Sullivan County as quickly as the heavy machinery could negotiate the windy roads and cut down wide paths through the trees. (Roads not designed for a heavy influx of trucks remain a problem. A local saying is that the county's largest road was designed by a slow snake; the second largest road by a fast snake.) The landscape is being transformed. Before 2006, you could drive over a crest in the road, and a picture postcard scene unfolded of hay bales in a field and a weathered gray silo with a silvery dome next to a red barn. Around the corner was a snapshot of rural poverty, a dilapidated trailer with a wooden porch that sagged precariously. Now a third image had joined the tableau: the straight brown strip of newly turned earth, cutting a straight line through a field or the woods, with lengths of green-wrapped steel pipe on wooden braces ready to be installed.

Bill Hart, a dairy farmer in his seventies, offered me a tour of these new pipelines on his low-to-the-ground, four-wheel-drive all-terrain vehicle. He drove like a teenager given the keys to the family car for the first time. He descended embankments with abandon. "I am going to take you up here and show you one of the things that bothers me to no end," he said, before gunning the ATV into a field, the engine grumbling so loudly that we had to yell at each other to be heard above the din. After negotiating a field of oats and rye, we drove out of the crops and down a two-foot cut into a brand-new pipeline right-of-way. Machines had come through and cleared out every hint of vegetation. Driving in this earthen artery, he paused to point out a sixty-foot tree right on the dividing line between the cut and the forest.

"See that nice cherry tree there? This one right here. Oh, that's a wonderful tree. But look what they did to the roots. That tree is going to die," he said. The machine that resculpted the land had sliced off part of the roots, leaving behind fleshy wood exposed. "Carl Driscoll used to own that land. He bought fertilizer from me for these trees. I don't know where he is buried, but I know he is turning in his grave." What bothers him, Hart said, is the waste. "They could do different, they do not need to cut eighty to ninety feet for a sixteen-inch pipeline," he said.

For the next hour, we sped around his and neighboring farms, cutting through fields and bumping along paths in the woods. He complained that gas companies weren't "conservative," a word he uses not in a political sense but to mean that they aren't intent on conserving. "Farmers are tighter than hell; we don't want to see anything wasted. You put your tomatoes on your plate, you clean up your plate. That is just the way farmers are." The gas companies are different. The pipeline rights-of-way will be kept clear so that gas companies can fly over them looking for any signs of leaks. "You will never use that land again; it will never be any good for anyone. That land is totally wasted," he said. "What is it going to harvest?" He has been farming and raising cows since the 1950s and has adapted to considerable changes in the dairy industry. But nothing prepared him for the changes brought by the gas industry: the wells and traffic and pipelines. He remembers his father signing two-year gas leases for a quarter an acre. "They didn't do anything. It was kind of free money," he said. I asked if anyone anticipated the magnitude of change in the past couple years. "No. None. I don't think anyone did."

On his tour, he pointed out not just who owned a field, but who owned it a generation or two ago. The land for him isn't merely a plot. It's a connection to the past and to his family. As we head back toward his house, he points out an acre lake with a couple small boats on the shore. "I built that after my first wife died. That first winter, just for something to do," he said. As he pulled up behind his house, he said he wanted to show me one more thing. He opened a garage bay, and inside was a meticulously restored 1969 Cadillac convertible, a shiny crimson car with a "Boss Hart" vanity plate. The car looked brand new. "Maybe I'm too much of a conservative," he said. "I don't know."

Through Hart's eyes, Sullivan County is being irrevocably remade by outsiders who don't share his reverence for the land. And to what end? The changes will provide an abundance of natural gas that will reshape the energy landscape just as surely as it is reshaping Sullivan County. Once the Marc 1 is connected, natural gas from

Sullivan County will become part of a surge of new gas heading toward coastal cities. The result, according to federal projections, will
be that gas—and to an extent renewable energy also—will displace
coal-fired power plants, sending somewhere between 11 percent and
22 percent of existing coal plants into retirement. Federal forecasters
say the availability of cheap natural gas, in fact, makes it considerably easier to introduce new clean-air standards and increase renewable power without sending power prices skyward. Marcellus gas
will help undo many of the changes wrought by a poorly conceived
energy policy in the late 1970s. Worried about declining supplies of
domestic natural gas, Congress outlawed all new gas-fired power
plants. In the nine years that the Powerplant and Industrial Fuel Use
Act of 1978 was law, the coal industry went on a building spree. One
of every five coal power plants in the country was built during this
window of time. Signing its repeal in 1987, President Ronald Reagan
pointed out that burning gas emits fewer pollutants than burning
coal. "As natural gas is a clean-burning fuel, restrictions inhibiting its
use have not been in the best interests of the environment," he said.

As Marcellus gas flows east to the coastal cities, forecasters expect it to keep power prices from rising too quickly. Electricity prices
are expected to rise slowly over the next couple decades, due to low
natural gas prices and improving energy efficiency. Sullivan County
will benefit from moderate electricity bills, but many residents may
never get a chance to use the gas to keep warm or boil water. The
towns are too small to justify construction of a local gas distribution
network. After Sullivan is drilled up, most residents will still rely on
heating oil and wood to keep warm in the winter.

There are other economic benefits to Sullivan County. The most
obvious is that landowners who leased in 2008 and 2009 received
thousands of dollars per acre, plus up to 25 percent of the royalties
from gas produced on their land. The influx of oil-field workers and

demand for oil-field services were having an impact. A company that refurbishes oil-field equipment had set up a shop in Sullivan County, as had another that provides sand for fracking operations. More than one hundred cars parked in the lot of the company installing the Marc 1. One longtime farmer and his wife told me quite happily that one of their grandsons had moved back because he got work driving a truck hauling water for frack jobs. In Dushore, the largest town in Sullivan County, a local dentist had taken out a loan to build the first new hotel in a century. Construction had barely begun when oil companies began inquiring about renting rooms for months at a time.

"This is the biggest—for lack of a better word—economic development in this county since lumbering," said Sullivan County commissioner Bob Getz, a garrulous man who ran for office after decades working for the state as an auditor. "We're just getting started here. It happened very quick, very fast," he said.

Down the street from the new hotel is the county's only auto dealership. Its parking lot is crammed with a couple dozen brand-new jumbo Ford F-250 pickups. They are selling briskly to oil-field workers. A rig worker, often with a high school education and a willingness to work long hours, can earn a nearly six-figure income. Other than drilling and fracking wells, this kind of paycheck is practically impossible to find in central Pennsylvania. But the hefty compensation is, in part, to compensate for the dangers of the work. In January 2011 a high-pressure hose was left unsecured on a drilling rig on the western side of the county. The hose, which delivered a specialized fluid to keep the drill bits operating smoothly, thrashed about and struck a thirty-year-old father of two young children in the head. He died by the time a helicopter airlifted him to a nearby hospital.

Development of the Marcellus Shale began slowly in late 2004. A small Fort Worth company called Range Resources decided to try a modern slick-water frack into the Marcellus Shale. The first well where it pumped a "Barnett-style" frack, using one million gallons of water along with a smaller quantity of chemicals, was southwest of Pittsburgh. Previously, the largest frack in the area had pushed sixty thousand gallons into the well. Early on, Range's biggest problem was that Pennsylvania was an outdoor museum of drilling equipment. To drill the experimental Marcellus well, called the Renz #1, Range wanted to use a giant spool of pipe called a coiled tubing unit that can be put into the well quicker than using straight segments that have to be screwed one into the next. There weren't any units in Pennsylvania, but Range found one in Kentucky. The driver hit a bridge abutment on his way north. When it arrived, there was no one with any experience using it. Range executives used the corporate jet to rush three operators up from Texas. Later, when Range needed an expert on drilling shale, it hired a consultant and told him to be in Pittsburgh within two days. He made plans to drive to nearby Pittsburg, Texas, before he realized the misunderstanding. When he arrived in Pennsylvania, he said it "was like stepping back in time a hundred years."

John Pinkerton, Range's CEO at the time, said it was a mistake to use available local drilling crews on the type of modern, complex wells he was drilling. "We had Appalachian equipment and Appalachian crew," he said. When they got a crew that understood modern fracks into southeastern Pennsylvania, they miscalculated the weather. "It was a bunch of guys from Texas" who didn't realize how cold it would get overnight, said Pinkerton. "Everything was freezing. It was just a disaster." Despite these missteps, Range knew that it was on to something. The Marcellus Shale had a lot of gas and, when the wells were built right, could be fracked. Range began leasing up acreage for $25 to $50 an acre. Within two years, Range went from having 38,000 acres under lease to 250,000 acres.

When the results of Range's first few wells became public, other companies started paying attention. Men parked on the roads near their wells, watching Range's wells through binoculars. "Aubrey has got people everywhere," Pinkerton said, referring to Chesapeake's CEO. Asked if they were spying, he replied, "Oh yeah. All the time." McClendon liked what his scouts saw. Chesapeake, in late 2005, spent $2.2 billion to buy a company with a lot of acreage in Appalachia and then unleashed its army of landmen. Chesapeake drilled its first horizontal well targeting the Marcellus in September 2007.

As usual, it was Chesapeake's giant appetite and willingness to spend freely that transformed the Marcellus Shale and the land above it. Acreage that could be had for $200 an acre spiked to $1,000 and kept rising until it peaked around $6,000. Pinkerton said Chesapeake's large wallet caused him—and any other company that was eying the Marcellus—to speed up and move faster. Landmen spread out across the state and leasing begat a drilling scramble. In 2007 there were 71 permits submitted to drill into the Marcellus Shale. By 2010, there were 3,316 permit requests. Equipment and labor were in short supply. New equipment was built and moved into the Marcellus, but it wasn't enough. Any rig that could be repurposed to drill a mile down was used, even though most of the equipment in Pennsylvania was intended for much shallower, lower-pressure wells.

Problems cropped up. In June 2010 EOG Resources was completing a well in Clearfield County, Pennsylvania, about one hundred miles west of Sullivan County. It was drilling out plugs inserted during the fracking operation when it lost control and spewed thirty-five thousand gallons of frack fluid and brine into the air. A state investigation turned up several problems. The crew—all local hires—had worked for twenty-four hours straight. Neither the crew, nor a more experienced supervisor, had up-to-date well control training. EOG failed to install redundant systems to prevent a blowout. There was only one barrier—a large set of rams at the surface that sealed around the pipe—and that was opened to replace rubber pads. Investigators could not determine who gave the order to open the rams, or if there had been a miscommunication, but concluded, "[T]his type of activity should never be allowed at any time." Instead of using a modern device to drill out the plugs, EOG used a 1940s rig configured to repair existing wells.

Incidents such as the one in Clearfield make some people in Sullivan County nervous. As the number of wells and pipelines and trucks grew, many local residents came to regret signing gas leases. What seemed like an easy way to supplement their meager incomes morphed into something much different. They had opened the door to these visitors, expecting a brief visit where they ended up pocketing these outsiders' money. But the folks from Oklahoma and Texas had arrived en masse and industrialized parts of the county to get at the gas. And evidence began to accumulate that the state didn't have a firm grasp on what was going on and how to keep Pennsylvanians safe. In April 2011, several years after the boom began, the state asked Marcellus drillers to stop sending their wastewater to all public water-treatment facilities. The industry complied and began recycling more of its wastewater, but by then, these public facilities had accepted millions of gallons of water. Some weren't equipped to clean up all the radioactive elements, such as radium, and pollutants, including strontium and barium. The water, after treatment, had been released into rivers—in some cases, upstream from public drinking water intakes.

Concerns about the state's water metastasized. Many worried that fracking would create cracks that allowed chemicals and gas to move up into shallow aquifers. It turned out to be drilling, not fracking, that caused the problems. In southern Bradford County, one of the largest documented cases of water contamination related to gas exploration occurred after a well had been drilled, before any fracking took place. A state investigation concluded that Chesapeake had failed to cement its wells adequately, allowing gas to leak from pipes into the groundwater. The company agreed to pay $900,000 in fines and payments to the state but never acknowledged publicly that it caused the problem. According to letters from Chesapeake to the state, the company said that its wells had intersected a fault in the earth that may have allowed gas to travel from the wellbore into a nearby beaver pond, where it bubbled to the surface. Chesapeake later agreed to pay three landowners $1.6 million to settle a lawsuit.

All of the early confirmed cases of escaping gas from the Marcellus Shale occurred in counties close to Sullivan County. What was going on in this part of Pennsylvania? There are a couple possible answers. There hadn't been many deep wells drilled in this area, so gas companies lacked a good grasp of the complex rock strata and didn't have time to wait to piece together the puzzle. The amount of geological upheaval in this area is remarkable. One geologist I spoke with pointed out that in Texas's oil-rich Permian Basin or Louisiana's Gulf Coast, rocks lay on one another in flat lines, like a stack of pancakes. The area around Sullivan County is more akin to a bowl of spaghetti. Some gas from the Marcellus, over millennia, had used the fractures and folds in the rock to migrate upward until it encountered shallow sections of sandstone and siltstone. The gas entered these rocks and saturated the pore spaces. When gas escapes from new, poorly cemented wells, it encounters gas-saturated rock strata. There is no place for it to go, like a rush-hour commuter trying to board a packed subway car. It keeps moving from rock strata to strata looking for an opening, until it finds a pathway upward toward a water aquifer or residential well. In other parts of the state, there are more low-pressure rocks for gas to seep into long before it finds a path up to the surface.

With several thousand wells drilled since the beginning of the Marcellus boom, and tens of thousands more planned, there have been some localized problems with water contamination. How many is hard to track. Pennsylvania reports how many wells have been drilled and how rapidly oil and gas production is growing. But it does not report how many complaints about gas in water have been called in, or how many homes can no longer use their water wells.

Chesapeake's former CEO Aubrey McClendon, in a speech in Philadelphia in 2011, conceded that poor well design had caused the "limited gas migration incidents." He pointed out that working with state officials, new cement and casing standards had been implemented. "Problem identified. Problem solved. That's how we do it in the natural gas industry." But that is not quite the case. In June 2012, a month before my trip to Sullivan County, Royal Dutch Shell was fracking several wells in nearby Tioga County when a resident driving down a dirt road a half mile away came upon a geyser of methane-laced water that shot twenty to thirty feet into the air. It was coming from an old 1932 well that had been plugged, abandoned, and forgotten. Shell mobilized a small army of engineers to figure out the problem and began venting several wells to relieve underground pressure. Members of a hunting club were barred from their cabin, and methane turned up in some other nearby wells. The state reported that Shell had to fix its cement in one of its new wells, suggesting that gas from the new well had escaped into the old well. Pennsylvania didn't even have the 1932 well in its list of abandoned wells. "The list is a work in progress. There are possibly thousands of those wells in Pennsylvania," said a spokesperson for the state's oil regulator.

The Shell well wasn't far from my parents' land. One afternoon I drove over. Even without an address, it was easy to find. A waitress at a local restaurant gave me turn-by-turn directions up a steep dirt road. Shell was flaring off gas from one of its recently drilled wells to bleed off pressure and prevent more gas from migrating out of the wellbore. There were a couple dozen pickups parked at the well and engineers trying to figure out how to undo the damage. On the next ridge of hills were several large wind turbines, a visual juxtaposition of two possible energy futures. Despite a small wind, the giant 230-foot-long sleek blades were motionless. They weren't generating any power.

I parked at a nearby farmhouse. A deeply tanned man wearing a baseball cap and loosely holding a large pair of pruning shears in his hand walked out of a field of four-foot-tall fir and spruce trees, planted in neat rows and destined for sale to landscapers across the East Coast. He introduced himself as Andy Proteus. He said that he leased the land his family owns to Shell for $100 an acre in 2007, not expecting much to happen. "The whole thing makes me nervous. It definitely makes me think twice before I lease my ground again," he said. Then he closed the door to his pickup truck. He was done talking about it. His trees weren't going to prune themselves.

There was a much smaller environmental headache on my parents' property. A few months after I visited the Farm, I received a panicked call from my father. Chesapeake had just called. While the company was boring a hole on the property to run a gas gathering line under a creek, some of the drilling fluid had escaped into the water. It wanted to take immediate action and needed the landowners' permission. I put him in touch with Wylie Norton, the Sullivan

County commissioner whose land backs up onto the creek a few hundred feet downstream of the leak. State inspectors were already aware of the problem. Within a couple days, Chesapeake had completed the cleanup. Fish in the creek had been killed, perhaps by toxins in the drilling mud used to lubricate the drill bit and carry off rock cuttings or maybe from being deprived of oxygen in the water. The state Department of Environmental Protection didn't issue any violations or fines.

Afterward, I asked Norton, who favors fracking, if the incident had changed his mind at all. "We always have some concerns about environmental damage to our area and special concern about drinking water safety," he replied in an email. A few sentences later, he vented his growing frustration about how acrimonious the debate over fracking had become. "Unfortunately, similar to the political climate in DC, there are two extreme groups debating the gas issues that refuse to objectively look at issues. I will always question the 'drill, baby, drill' folks as well as the 'Frac no' people," he wrote. "Are the gas companies looking for profits? Sure. Do they have absolutely no regard for the environment? No. Will there be problems and accidents when there is this large scale of an operation? Sure!"

While landowners each made individual decisions about leasing their land, the future of the county will be determined by the combined impact of hundreds of wells. In early 2011, two prominent geoscientists—one a booster and the other a critic—met for a public debate in the Sullivan County High School auditorium. The topic was "The Cumulative Environmental Effects of Gas Drilling." It might as well have been called "Fracking: Is It Worth It?"

Terry Engelder spoke first. A Penn State University professor of geosciences and fracture mechanics, he grew up a couple counties away from Sullivan across the state line in western New York and spent his career focused on rocks and their natural fractures. He has good reason to hold a grudge against oil companies. As a teenager, the ring finger on his right hand was crushed between two pipes while he worked as a laborer in an aging Pennsylvania oil field. But he has become an unabashed supporter of shale development. His interest in the Marcellus Shale was piqued when he was asked a few years earlier how much gas was in the rocks. Stumped, he made some rough calculations on a piece of scrap paper he pulled out of the recycling bin under his desk. The answer he came up with was as much as 516 trillion cubic feet. "I didn't believe it," he said. He called up a colleague and asked him to do the same calculation. He too came up with roughly the same figure. As the size of the gas field sunk in, he said it was like an "out-of-body experience."

On the stage at the high school, Engelder, a slender man with deep creases in his face and a bulbous nose, wore a blue ski sweater and blue jeans. He described himself to the skeptical audience as a "shale gas enthusiast." Quoting President John F. Kennedy's famous call for sacrifice—"Ask not what your country can do for you"—he talked about the importance of the shale for the future of the United States. "The people of Sullivan County are being asked to sacrifice," he said. "The sacrifice, of course, is that this is an industry that can't come in without leaving some scars behind. It is necessary to build drill pads, it is necessary to run pipelines and whatnot to exploit this energy. The question comes down to one of how does America move forward into the future? What is its energy?" He said he had an abiding "faith in American industry's ability to do it right." There will be some mistakes along the way, he admitted. He felt bad for people who lived where the industry had drilled bad wells. But shale development is a force for good, he said, and there must be some "necessary sacrifice."

Energy, he said, is joined to economic growth. The American lifestyle depends on energy. If we turn our backs on shale gas, we'd be eliminating 20 percent of the energy Americans need to live our lives. Taking this point to an extreme, he recommended the audience read a couple postapocalyptic books, including Cormac McCarthy's fictional *The Road*. If the United States says no to new energy development, the result would be economic ruin. "In terms of environmental devastation, nothing is more severe than total economic collapse," he said. The audience squirmed.

Anthony Ingraffea spoke next. He is a professor of civil and environmental engineering at Cornell University, an Ivy League college where students can major in comparative literature but not petroleum engineering. He is a prominent critic of fracking and co-authored a controversial paper that suggested so much methane gas leaked from wells during their lifetime that shale gas might produce more greenhouse gases than oil or even coal. The paper's conclusions have been attacked by a number of subsequent researchers, some of whom were funded by the industry and others who were independent. He was dressed in a suit and tie and had a shock of white hair above a round, fleshy face.

"Tonight we are here to talk about cumulative environmental impact. Cumulative environmental impact doesn't mean what goes wrong when one cement job fails. Cumulative environmental impact doesn't mean what goes wrong when one set of wells on one pad goes sour. Cumulative environmental impact means what is happening in a county with fifty pads? What is happening in a region like Pennsylvania with ten thousand pads?

"Or the biggest question of all, and the one for which we have generational responsibility, is this one: that is the global environmental impact, and that refers to the question of should we even do it," he said, drawing murmurs of agreement from the crowd. "How do we get from where we are today, which is not sustainable even with shale gas that is a nonrenewable resource, and it won't get us past our grandchildren, our grandchildren's generation? How are we going to get to that golden era when we stop kicking the can down the road to our kids and grandkids and suck it up—or as they say nowadays, man up—and solve the damn problem now?"

John Trallo moderated the three-hour debate. It helped crystallize his thinking about fracking. He moved up to Sullivan County from Philadelphia in 2001, not long after his wife died of breast cancer. He doesn't consider himself an environmentalist, but his wife made him promise to move somewhere away from chemicals to raise their son. He chose Sullivan County because he remembered camping there in the 1970s. A music teacher and sometime recording session guitarist, he found clients who wanted music lessons or to record a demo. One of his students made it to the top five on the TV talent show *American Idol* in 2010.

He turned down offers to lease his 1934 house, which sits on eight-tenths of an acre in the village of Sonestown. "They made me an offer to industrialize my property. I said, 'That's not why I moved here, and I'm not interested,' " he said. The industry settled all around him. Down the street, a dilapidated schoolhouse was fixed up and turned into housing for twenty-eight oil-field workers. During the summer, he keeps his windows closed and runs his air-conditioning to keep out the smell of diesel wafting from a nearby two-lane road that has become the primary way for trucks to get into Sullivan County. He finally offered to sell his house to an energy company, but didn't get any takers.

Eighteen months after the Engelder-Ingraffea debate, he said his most vivid recollection of the night was his reaction to Engelder calling on Sullivan County to make a sacrifice for the good of the country. "No one asked me to make that sacrifice, and I think I've sacrificed enough," he said. "It is very easy for those not living here to ask us to make a sacrifice."

On my drive out of Sullivan County, back to Philadelphia, Trallo stuck in my mind. Most of the farmers I met signed a lease and agreed to take bonus checks drawn from Chesapeake's treasury in exchange for letting wells be drilled. Like the Reibsons, some ended up feeling steamrolled by the changes. Trallo never signed a lease or took any money. He chose Sullivan County as a bucolic place to live and raise his son. The arrival of fracking blindsided him.

I also thought about what a longtime resident had told me. One afternoon I ate lunch with the Shoemakers. The father, "Doc," was a longtime veterinarian known around the county for his skillful care for heifers. More recently, residents got a chuckle when, in his nineties, he chased a black bear near his house in town to get a good photograph. His son John owned and edited the *Sullivan County Review*, the local weekly paper. Both embraced the jobs created by Marcellus drilling and saw it as a much-needed boost to the local economy. Over sandwiches at Pam's Restaurant, John said he was having trouble understanding fracking opponents. All the gas being drilled in rural Pennsylvania was headed to the big cities near the coasts. The city dwellers were getting all of the benefits of this new source of energy, while the locals were the ones who had to wait at the county's only traffic light behind large eighteen-wheelers carrying fracking equipment and watched as their landscape was crisscrossed by pipelines. Shoemaker said most locals were willing to make the sacrifice because of the economic opportunity being created. The strongest antifracking sentiment came from the cities, he said, where people weren't being asked to give up anything. "Our gas from Sullivan will go from here to New York and Philadelphia and lower the electricity rates of the people protesting against us," he said. It was an irony he had trouble swallowing.

I thought about his point as the miles rolled by. Philadelphians didn't have to choose whether to lease their land. The Marcellus formation ends before the city's western suburban sprawl begins. But they did use a lot of energy and expected it to be available to keep them warm. Consuming energy while protesting against energy production seems hypocritical. What were the growing antifracking protests against? Who was right?

11

BLESSINGS OF THE POPE

It was a beautiful fall day for a protest. A few hundred people marched through downtown Philadelphia chanting, "Hey, hey, ho, ho, hydrofracking's got to go." By the time they arrived outside the Pennsylvania Convention Center, the crowd had swelled. Many carried "Ban Fracking Now" posters. One woman wore a "What the Frack" T-shirt. A Lutheran pastor in full vestment addressed the crowd, a rabbi blew a ceremonial ram's horn as a "wake-up call." Filmmaker Josh Fox, in his trademark baseball hat and thick-framed glasses, told the crowd that "fracking is a disaster unfolding across Pennsylvania."

Inside the convention center, more than a thousand energy executives were attending a conference on the Marcellus Shale. The keynote speaker, Aubrey McClendon, took the stage and called the protesters naive. "Their real game plan is to use political pressure to force Americans to pay exorbitant energy costs for the so-called 'green' fuel sources that they prefer," he said. But what would happen if we followed their lead? he asked. Wind and solar power can't power the US economy. If we ban fracking, natural gas prices would skyrocket. Crops that require natural-gas-based fertilizers would cost more. Homes and business and factories would lose their heat and power.

"What a great vision of the future! We're cold, it's dark, and we're hungry. I have no interest in turning the clock back to the Dark Ages as our opponents do," he said. It was typical McClendon: a provocative, in-your-face attack. To the protesters outside, he was the face of fossil fuels and a man bent on destroying the environment.

Carl Pope doesn't agree. "I think Aubrey McClendon will undoubtedly turn out to be one of the major contributors to giving the world a shot at protecting the climate," he told me in an interview. Pope ran the Sierra Club, the nation's largest environmental group, for eighteen years. The Sierra Club, founded in 1892, has been at the forefront of fights to preserve wilderness and clean up the environment. During his tenure, Pope began to believe that protecting the environment in the twenty-first century meant fighting climate change.

"The function of the environmental movement is to enable local people to defend local places from immediate threats," he said. "But the science is telling us these places aren't even going to be here. The Arctic National Wildlife Refuge is going to be underwater. So what are we talking about? Forget protecting it, let's just stabilize the climate. Without a stable climate, the whole idea of protecting places with ecosystems and critters becomes impossible."

In a perfect world, he knew fracking had an environmental impact. But the world that Pope saw wasn't perfect. It was facing a dire threat. To stop the planet from heating up meant stopping new coal-fired power plants from being built and shutting down existing coal plants. In that fight, Pope and McClendon became improbable allies. The natural gas that Chesapeake and others were finding became an invaluable tool. "I thought, and still think, that Aubrey was a perfectly reasonable environmentalist, but that wasn't what motivated him. What motivated him was he needed markets for his fuel," said Pope.

The unusual partnership between Pope and McClendon began in 2006. Pope had recently guided the Sierra Club toward a greater emphasis on energy and shutting down coal plants. A couple friends had suggested that he meet with McClendon, who they said was also interested in fighting coal. The men had different motivations but a common goal. They met first in Connecticut, near an energy conference where McClendon was speaking. The energy executive began explaining to the environmental leader about fracking and how the once-scarce natural gas was about to become more abundant. McClendon explained that a lot more gas was going to be available to replace coal to make electricity.

Hearing this was a revelation to Pope. "It is going to run a few power plants and heat a bunch of homes—that was what we thought natural gas was for, and that was fine," he recalled. McClendon continued his pitch, explaining that a kilowatt-hour of electricity generated from gas released half as much carbon into the atmosphere as the same kilowatt-hour from coal. Pope was intrigued. He had never heard of fracking before and shared the conventional wisdom, at the time, that gas would remain scarce. Toward the end of 2006, the two men soon met again, in Washington, and then a third time, in Oklahoma City.

When Pope first met McClendon, each man was at a turning point in his life and career. McClendon was still on his way to enormous wealth and success that was built precariously on debt and the most volatile of commodities. If Chesapeake and the rest of the industry overshot and produced more gas than needed, prices could collapse. One way to prevent this glut was to get more industries, power plants, and people to burn natural gas—even if this meant denigrating coal. Pope was trying to steer the Sierra Club, the nation's largest, most influential environmental group, away from its historic focus on protecting places and endangered creatures. He wanted to focus on the climate. Enemy number one was coal. Traveling very different paths, Pope and McClendon had arrived at the same place.

Before they met, McClendon had already begun to develop his own plan of attack against coal. The first salvo was unleashed on February 4, 2007.

Ken Kramer was at home in Austin doing what he usually did on Sunday morning: reading the newspaper while sipping his coffee. For nearly two decades, Kramer had the often thankless job of being director of the Lone Star chapter of the Sierra Club. His years at the helm of the state's Sierra Club made him the state's unofficial top environmentalist. He knew just about everyone in Texas who fought to save state parkland, protect endangered species, and lower industrial emissions.

As he flipped through the pages of the *Austin American-Statesman*, he came across a full-page ad that caught his attention. Staring out from the page was an enormous close-up photograph of a beautiful young woman with azure eyes. Her face was covered with dark smudges, evoking midcentury images of sooty coal miners. "Face it. Coal is filthy," the ad read. "Texas needs clean skies. Not black skies. Stop the filthy coal plants."

"Where did this come from?" he asked himself. Then another thought, "This cost a hell of a lot of money." Who had the money to pay for this giant ad? Certainly not the Sierra Club or any other environmental group that came to mind. He scanned the ad, and his eyes landed on the name of the group at the bottom of the page. What was the Texas Clean Sky Coalition? The ad encouraged people to attend an anticoal rally the following Sunday that Kramer had organized. It was as if he had just learned he had a rich, generous uncle, he thought. The uncle's name was Aubrey McClendon.

The rally protested a proposal to build several new coal plants wending its way through regulatory agencies. Texas's largest electricity utility, TXU Energy, was a century-old pillar of the state's business establishment. It had been around, in one form or another, since the first forty streetlamps were installed in downtown Dallas. In the early 2000s, the company had made an ill-conceived investment in European power plants. The move had gone poorly. A new chief executive was hired to clean up TXU's balance sheet and restore it to profitability. After studying the market, John Wilder came up with a simple back-to-basics plan: burn coal and generate electricity. In April 2006 he unveiled his ambitious plan. TXU would spend $10 billion to build eleven new coal power plants. The protesters wanted to stop him.

Sitting in his one-story Austin home, Kramer wondered about the Texas Clean Sky Coalition. He figured it was one of the new ad hoc groups that had popped up to fight TXU's coal plants. Whoever it was had a big checkbook, he thought.

The reason Kramer didn't know anything about the Texas Clean Sky Coalition was that it had come into existence only a couple weeks earlier. The name was misleading. The coalition was one company—Chesapeake Energy—and the unfolding campaign was underwritten by Aubrey McClendon. He had decided to get into a scrap with coal. Over the next few days, Kramer got calls from Sierra Club activists in other parts of Texas. The ads were running in Dallas and Houston and elsewhere. The ads varied—there was a young white girl, a Hispanic male, an older white woman—but the message was consistent. Coal was deadly. "Live longer. Live better," the ads stated. "No new coal plants."

It was a stunningly effective campaign and marked a significant turning point in US energy history. Energy companies can be fierce competitors against one another in the marketplace, but there was a gentlemen's agreement that they didn't attack one another publicly. They might battle to win allies on state utility commissions and in Congress, but these skirmishes took place behind closed doors or in obscure regulatory meetings in state capitals. McClendon hadn't just disregarded this deal, he had torn it up into a thousand pieces, doused them in lighter fluid, and lit them on fire. The head of the National Mining Association, the Big Coal lobby in Washington, later sniffed that a McClendon-backed attack on coal "marks a disturbing departure from the understanding we tacitly share in the energy sector to avoid denigrating competing fuels." Perhaps the lobbyist didn't understand. That was McClendon's goal.

Soon after the ads appeared, Chesapeake acknowledged its role in funding the campaign. "It's simply that we think reduced emissions," said Tom Price Jr., one of McClendon's closest advisers, "is a good thing." There was another benefit to stopping TXU's coal plants that he didn't mention.

It costs a lot of money to build a new coal plant. To recoup those costs, coal plants run for decades. Once built, TXU's new coal burners would be a major force in the giant Texas power market for generations. They would elbow competitors off the grid with their cheap power. Texas, with its large population and its power-hungry refineries and petrochemical plants, uses more electricity than any other state. The Texas power grid was a big market for natural gas. The amount of electricity generated by natural gas in Texas was nearly twice as much as in California. Texas, by itself, generated more power from gas than thirty-eight other states *combined*. If TXU built the eleven new power plants, demand for natural gas would go down, and so would prices. And not just in Texas; the impact would be nationwide.

This didn't sit well with McClendon. Chesapeake had been leasing up gas fields and promised investors it would continue ramping up production. McClendon also saw, before nearly anyone else, that an abundance of natural gas was about to hit US markets, even if he failed to fully grasp how much. Chesapeake had bet its future on natural gas, and TXU wanted to cannibalize one of his biggest markets. As for the environmental benefits of burning gas instead of coal? McClendon saw the greatest threat from a warming planet was that winters would get warmer, cutting demand for natural gas to heat homes. He didn't view climate change as an existential threat. It was a business opportunity. McClendon, arguably, figured out how to profit from bizarre weather trends that *New York Times* columnist Thomas Friedman later termed "global weirding." Extreme cold or hot snaps cause volatility in natural gas prices. McClendon used that volatility—"We crave volatility," he once said—to generate billions of dollars for Chesapeake by trading gas futures. Even as he profited from weather gyrations—and employed his own two-man weather team—he said he wasn't even sure if the science predicting climate change was right. But he realized that the growing anxiety about climate change created an opportunity for him and for Chesapeake's growing reserves of natural gas.

TXU had strong support for its Texas coal plants. The day it announced its plans, its stock jumped nearly 8 percent, signaling that investors were pleased. Texas's governor, Rick Perry, also backed the plan. Perry had appointed the regulators who would vote on whether to issue the needed air permits. McClendon didn't have many options except declaring war on coal, which is what he did. His anticoal campaign would prove enormously successful, but only partly because of his involvement.

TXU was facing resistance from environmental groups, as well as from the mayors of Dallas and Houston. Several prominent Dallas businessmen, including real estate magnate Trammell S. Crow, came out against the plan. They asked TXU to scale back its coal plans or at least slow them down, but were rebuffed. TXU's political strategists considered this opposition manageable. But there was another, thornier problem with Wilder's coal dream. TXU's plan to haul millions of tons of coal from Wyoming was in trouble. It wasn't clear if the railroads could deliver all the coal that was needed. And the shipments they could provide would be expensive. The arithmetic of Wilder's coal plan wasn't adding up. He needed to scale back. Internal doubts about the plan's feasibility grew. But TXU and its ally in the governor's mansion had expended a lot of political capital to get the permits approved. The plan was to get state approval for the eleven new coal plants and then retreat.

Some on Wall Street sensed this weakness and began circling the wounded TXU. A couple months before the "Coal Is Filthy" campaign began, two large private equity firms, Kohlberg Kravis Roberts & Co. and Texas Pacific Group, approached Wilder about buying the company. During preliminary discussions, the Wall Street firms pressed TXU for more information about its coal plans. Then McClendon's ads began running. Already weakened by its railroad problem, TXU surrendered. On February 9, 2007, less than a week after the first ads appeared, TXU's lawyers agreed to the buyout. Negotiations would continue in secret for a couple more weeks, but the deal was basically done.

Two days after the preliminary agreement was struck privately, about two thousand people rallied in Austin at the capitol to protest the coal plants. With Chesapeake's "Coal Is Filthy" ads running in every major Texas newspaper, turnout was boosted for what was Texas's largest environmental rally in years. Some protesters wore "Face It. Coal Is Filthy" T-shirts. Others held "Vote No on Coal" placards. A woman in a bright floral dress, with a large headdress in the shape of the sun, with streamers attached, danced around a man with a smudged face wearing a large, black smokestack.

When the TXU buyout deal was announced in late February, the new owners capitulated to the anticoal opponents. They would seek to build only three new coal plants, "preventing fifty-six million tons of annual carbon emissions," according to a press release. The death of TXU's grand vision to power Texas well into the twenty-first century with coal plants came about at least partly because the railroads had the stronger negotiating position and planned to force TXU to pay a steep price for the coal it wanted. But McClendon's "Coal Is Filthy" campaign also played an important role. TXU and the private equity firms that wanted to buy it realized they were facing a fight not just against mayors and environmental groups that held little political power in Texas and had little money. They also faced an opponent willing to break with convention and drag TXU's reputation through the mud. McClendon had sent a message to all concerned. He had a lot of natural gas, he was prepared to fight, and he was willing to spend money to accomplish his goals. The Los Angeles–based advertising agency that put together the ads said at the time that the "Coal Is Filthy" campaign cost "north of a million dollars."

McClendon's easy victory against the Texas coal plants emboldened him. Later in 2007, Chesapeake decided to fight a new coal plant in its home state. "Coal is the wrong answer for Oklahoma today," McClendon told a regulatory body considering the plant. "We should not be importing trainloads of dirty coal while at the same time export clean-burning natural gas. Coal is a contributor to air pollution, water pollution, and it's the wrong solution for Oklahoma's growing energy needs." In a letter to a newspaper in Tulsa, he stepped up his attack. "Coal is simply on the wrong side of history— it is a twentieth-century technology that is completely unsuitable for meeting the new energy and environmental challenges of the 21st century," he wrote. "Oklahoma-produced, clean-burning natural gas is the fuel of the future." Before the end of the year, he had won this fight also. The state denied the power companies' request to charge ratepayers $1.8 billion to build the coal plant. Three years later, natural gas passed coal as the largest source of power generation in Oklahoma. Two years after that, for a month in the spring of 2012, as much electricity was made in the United States by burning natural gas as with coal for the first time ever.

The Texas coal skirmish marked the beginning of a larger battle for the future of US energy. Eventually the maelstrom that McClendon started would suck in renewable energy companies and threaten the survival of the nation's oldest and most prominent environmental group. It caused environmentalists to rethink long-held positions on the future of the planet's climate and whether allowing the industrialization of forests would do more good than harm. Fear that fracking would lead to widespread water contamination spread across the Northeast and became a rallying cry for opponents, but the rhetoric never matched reality. Within a couple years, the largest grassroots environmental movement in a generation rose up against McClendon's gascentric vision of the future. Chesapeake changed from a small, obscure Oklahoma company into a boogeyman. But the change unleashed by the gas industry's fracking also helped accomplish something almost unprecedented. All that natural gas cut

US carbon emissions. The reverberations from this epochal shift can still be felt, if not understood.

McClendon wasn't content to win a couple regional fights. As usual, he wanted more. He wanted—and needed—to keep drilling and producing gas. If supply was going to rise, he wanted to make sure that demand rose as well. Otherwise prices would fall. That's Economics 101: basic supply and demand. In the late summer of 2007, a plan came together in his head. If he could convince Texas and Oklahoma that natural gas was a better fuel than coal, why stop there? Why not preach the gospel of gas on a national basis? Since there was no national energy policy, McClendon set out, essentially, to create one. Concern about global warming was rising, and natural gas, while not carbon-free, could generate a lot more kilowatts of power for each ton of carbon emitted than coal could. Asked why he fought the Oklahoma coal plant—whether it was blatant self-interest or something loftier—he said both were true. "What's the matter with self-interest? From my perspective, I have a better product than coal," he said in a 2007 interview. Years later, he called coal "an anti-quated fuel contributing to serious health and pollution issues . . . I was determined to speak when others didn't seem willing to do so."

To deliver this message, he needed allies that could reach people not inclined to listen to an energy executive. He wanted natural gas to be sanctioned by an esteemed environmentalist. As usual, McClendon got what he wanted. After several meetings with the Sierra Club's Pope, he received a papal blessing, of sorts, for his chosen fuel. Within a year of their first meeting, Pope began promoting the environmental benefits of natural gas. It was a strange alliance.

The partnership began with money. In October 2007 McClendon made the first of $56 million in anonymous pledges to the Sierra Club. It was one of the largest donations in the history of the organization. At the same time, he also created the American Clean

Skies Foundation, a national outgrowth of the Texas Clean Sky Coalition. McClendon was the chairman of the foundation and its main financier. (Tom Price and Ralph Eads served on the board.) The Chesapeake CEO pursued a two-pronged strategy. He could use the American Clean Skies Foundation to promote the benefits of US natural gas across the country. It was a domestic energy source and has half the greenhouse gas emissions as burning coal, the group trumpeted. Plus, it doesn't emit any of the other toxins that coal produces—the nitrogen oxide and sulfur dioxide—associated with an increased incidence of asthma. Privately, he provided the money for the Sierra Club to expand an ongoing anticoal campaign. The Sierra Club's effort had focused on coal plants in the Midwest. With McClendon's money, it went national.

The first inkling that Carl Pope had befriended the fossil-fuel industry came in early 2008, in the *Oil and Gas Investor*, a small glossy magazine that delivered monthly doses of glowing executive profiles and insights into which companies were drilling where. In an interview, Pope talked about his willingness to embrace gas. "Among the fossil fuels, natural gas is at the top," he said. "There's a lot of opportunity—people in the natural gas industry tell me—to produce more natural gas domestically by using new technologies, and we're in favor of that."

The alliance was a marriage of convenience, not ideology. Pope, the environmentalist, viewed climate change induced by massive increases in carbon in the atmosphere as a global threat. He expounded on this theme in his monthly column in the Sierra Club's magazine. If the warming of the Earth isn't stopped, he argued, all previous club victories such as preserving wild places and protecting wildlife would be pointless. McClendon, the energy executive, was a capitalist who believed in natural gas, the down-home religion of Oklahoma. He was engaged in a fuel fight, not an environmental fight. Pope wanted to eradicate coal to cut carbon emissions. Mc-Clendon wanted to beat down coal because it would create a larger market for natural gas. They were fighting the same fight, but for different reasons. Still, the two men seem to have forged a genuine bond. "I'm a big fan of Carl," McClendon said years later.

At the time, McClendon was well on his way to becoming a billionaire. His personal fortune was tied mostly to Chesapeake, which was tethered to natural gas. When gas prices rose, so did Chesapeake's share price and McClendon's wealth. Pope, meanwhile, was struggling to keep the Sierra Club's coffers full. Following the terrorist attacks of 2001, the nation's attention moved away from environmental causes. Pope said later that after 9/11, "the Sierra Club and all other membership organizations started getting less and less individual donations. So we became more reliant on money that came with strings. That's the reality of the world." Even with McClendon's donations, the Sierra Club's finances weren't robust. Pope would soon cut funds for local chapters and lay off staff.

The two men also shared certain similarities. Both had run their organizations for years. McClendon had been at the helm of Chesapeake since the 1980s and ran it as if it were his personal fiefdom. He was a visionary and expected people to follow his lead. All critical decisions and many smaller decisions at Chesapeake ran through him. He set the tone and expected everyone else to follow. Pope had been executive director of the Sierra Club since 1992. It is no exaggeration that for many environmentally minded Americans, Pope set the tone for what to think about the complex issues of climate change, energy, and conservation. His vision and energy helped revitalize the environmental group and expand its membership. Pope was brilliant and possessed a seemingly boundless intellect and energy, much like McClendon. But by this point, he was also beginning to keep his board of directors in the dark about certain critical decisions. He never told them that McClendon had become a major donor.

To keep the McClendon donation hidden, Pope engaged in linguistic gymnastics. In 2008, asked point-blank at a speech in Cleveland about rumors that the Sierra Club owned an interest in oil wells, Pope replied, "We will not take money from oil companies. We have no connection whatsoever. . . . That is just a rumor." At the time, Chesapeake was producing thirty-one thousand barrels of oil a day. The donations came from McClendon himself. Technically, what Pope said was true, even if it doesn't pass the smell test.

The identity of Sierra Club's secret donor was revealed to Michael Brune shortly before he took over from Carl Pope as director of the Sierra Club in March 2010. He had previously run the Rainforest Action Network—"Environmentalism with Teeth" was its slogan—and he was an activist at heart. He once took over the PA system at a Home Depot in Atlanta and announced, "Attention shoppers. Aisle ten features wood logged from Indonesia, destroying the lives of thousands of indigenous people." Another message that day: "Attention shoppers. Thank you for buying ancient redwoods from California." The police escorted him out. Hired by the Sierra Club, he was preparing to begin his new job when a senior Sierra Club executive said that he should take a trip to meet the group's largest donor in his first few weeks. Brune asked who that was and where he would be going. The executive said she couldn't tell him. Brune guessed it was T. Boone Pickens. No, was the response, but he was warm. His second guess was Aubrey McClendon.

It's not unusual for nonprofit organizations to get anonymous donations. Donors seek anonymity for all sorts of reasons. Some don't want to be bombarded with pitches from other groups. Others seek to avoid calling attention to themselves, preferring to let the work they support speak for itself. Generally, someone on the board of directors would be briefed about a large anonymous donation. But McClendon's multimillion-dollar donations to the Sierra Club were a guarded secret, with only a tiny group aware that the nation's leading environmental group was receiving funding from one of the nation's largest fossil-fuel producers. Pope never told the board of directors.

When Brune took over the Sierra Club, his first major initiative presented to the board was to recommend snubbing the group's largest funder, at a time when the group was struggling for money. At a weekend retreat of the board of directors in August 2010, he broke the news of McClendon's donations to the all-volunteer board, a group of activists, lawyers, and scientists. In addition to informing them about the $26 million already received, he also suggested that they turn down another $30 million in promised pledges. (The donation amounted to about 12 percent of the Sierra Club's contributions between 2008 and 2010.) "It was a shocking time," board member Jeremy Doochin recalled. "And it was a very tough decision. You had a lot of money being put to a good cause, shutting down coal-fired power plants." The board discussed it for a couple hours and agreed to Brune's recommendation.

Brune said it was the right choice: "We needed to be independent from companies and industries whose practices we needed to change. We can't simultaneously accept any money, much less millions of dollars, from a company that clearly had environmental problems in an industry that was dramatically in need of reform." The board agreed.

McClendon's funding of the Sierra Club had remained a secret for two and a half years. During that period, Carl Pope and Aubrey McClendon stumped for gas. They appeared at industry conferences and visited with members of congressional staffers. Their message was that gas was now abundant. It was no longer a scarce resource and could be relied on as never before. Gas wasn't perfect, they argued, but it was a bridge fuel toward a low-carbon future. Using more gas for power generation could serve two goals. First, it reduced the need to burn the coal, which released twice as much carbon into the atmosphere. Second, natural gas power plants were more flexible than coal or nuclear plants. They could be switched on and off, and could run at a quarter capacity. They were an excellent complement for renewable power sources such as wind. When the currents blowing through a wind farm died down, operators of power grids needed another power source to fire up rapidly to keep the system balanced. Modern natural gas plants could do that; coal couldn't.

This argument, however, was out of step with the Sierra Club's own thinking. A couple years earlier, facing increasing evidence that climate change was accelerating, Pope had led the group through a several-month process of drafting a seventeen-page policy statement of its beliefs about energy and the environment. Using less energy by improving fuel mileage and improving weatherization was the best option. Wind power and solar panels on rooftops were the "preferred" sources of energy. Acceptable energy included small hydroelectric dams and developing ways to turn inedible plants such as switchgrass into ethanol. Finally, there were traditional energy sources that could be used during the transition to a low-carbon future. The Sierra Club opposed energy from burning coal, crude from Canada's oil sands, or the development of new natural gas fields in the United States. Pope told me that when the policy was drafted, neither he nor anyone else in the Sierra Club thought natural gas was abundant enough to play a big role.

Putting together this policy had been a monumental task. Changing it would be equally trying. Carl Pope opted to depart from the organization's stated priorities. He took McClendon's money and began publicly stumping with the gas executive, a man whose every statement to Wall Street boasted about Chesapeake's plans to drill thousands of new wells and develop new natural gas fields.

Carl Pope's new thinking about natural gas soon came to a head with Sierra Club activists in the Northeast. Early drilling of Marcellus Shale wells in southwestern Pennsylvania had proven fruitful. The industry set its sights on leasing land above the rock formation from West Virginia all the way into upstate New York. Of these states, New York turned out to be most resistant. In 2008 a bill was introduced in the state legislature to modernize its drilling regulations and permit the newfangled horizontal shale wells without a lengthy review or public hearing for each one.

As the bill made its way to the governor's desk, Roger Downs tried to figure out what it all meant and what he should do about it. Downs was one of two paid staffers for the Sierra Club's New York chapter and was its main lobbyist in Albany. Soon after he was hired in March 2008, he began getting calls from members visited by gas industry landmen asking them to sign leases. He was also hearing from grassroots groups in New York and Pennsylvania with horror stories about how intrusive natural gas drilling was.

Downs decided to see for himself what fracking would mean in New York. On a summer day in June 2009, he set out from Albany and drove southwest for more than an hour. He parked at a cemetery on the outskirts of the small village of Maryland. He grabbed binoculars, a notepad, and a handheld GPS. Cutting across private property, he hiked up a hill toward where the Ross #1 well was to be drilled.

At the time, the Ross #1 well was nothing more than a permit in a New York State office. Gastem USA, a unit of the Canadian oil and gas explorer Gastem, had applied for permission to drill and frack a well in Otsego County to see if there was gas in the Marcellus Shale in upstate New York. The state government was still debating whether to allow the kind of large-scale hydraulic frack jobs that were becoming commonplace in Pennsylvania. Under existing rules, Gastem could drill and frack it with no more than eighty thousand gallons of liquid. It would be a tiny frack, one twenty-fifth the size of a typical well. But Gastem was intent on drilling it. For Downs, the Ross #1 had a couple attractive qualities. It was not too far from his Albany office, and the permit listed the longitude and latitude where it was to be drilled. What better way to learn about fracking than to drop in on the well a few times over the summer and fall?

As Downs walked up the hill, the sound of faraway cars faded. He was in an unperturbed hemlock forest. He crossed a couple overgrown logging roads, but otherwise he had left civilization behind and was hiking through an intact ecosystem. He stopped to check his GPS, made some slight course corrections, and pushed on up the hill. After he had covered about a mile or so, he found a single stake in the ground with a Day-Glo orange tag on which had been written "Ross 1."

He stood there and absorbed it all. "It was this undeveloped, unfragmented forest," he said. "And then the context hit me. The area could survive this one well. But Gastem had forty thousand contiguous acres." How many wells would be drilled in this area? As he looked around, he found a large wetland not four hundred feet north of the future drilling pad. "I was shocked. I couldn't believe how close it was."

Downs was a wetlands biologist by training. He had spent years studying these ecosystems. At a previous job, he spent nearly a decade repopulating the Susquehanna River with shad, coming up with a plan that included a fish elevator around a dam. He did a quick inventory of the wetlands near the well site, taking samples of the grassy sedges and jotting down notes. By the time he returned to his car a couple hours later, he was troubled. Drilling here in this forest? It didn't make any sense. By the time he returned to Albany, he was ready to begin fighting fracking in earnest.

But there was a problem. As the Sierra Club's lobbyist in New York's capital, he knew the national Sierra Club's leader was in favor of natural gas and fracking as well. Carl Pope was speaking in favor of gas alongside Aubrey McClendon. And there was a "one club" philosophy. The Sierra Club was strongest when it spoke with one voice, and that voice was Carl Pope speaking in favor of gas drilling. Over the past year, as New York had debated changing its rules to allow modern gas drilling, the state chapter had staked out a lonely position. It didn't oppose the bill that updated the state's rules, but urged a thorough environmental review that would delay gas drilling for years. A Chesapeake lobbyist buttonholed him to ask what he was doing. How could you be fighting this with us giving you so much money? Downs recalled being asked. The encounter disturbed him but also seemed to confirm a hunch. The national Sierra Club had been downsizing and letting go of staffers. Positions that opened were in the "Beyond Coal" campaign. Downs had connected some dots. The organization had money for the anticoal campaign, and Carl Pope was palling around with natural gas companies. He suspected that the national Sierra Club, based in faraway San Francisco, had gotten a little too cozy with the companies who now wanted to drill a lot of wells in New York State. "It was very uncomfortable," said Downs. "In the Sierra Club, we were very divided. I think everyone was well intended. The national club took the thirty-thousand-foot approach to energy policy, whereas the chapter was looking at real experiences on the ground."

Downs returned to the Ross #1 several times over the next few months. Eventually Gastem invited him to observe the frack job in November. He spent ten hours on the pad, watching a crew frack the well and then another crew arrive to extract the water. "I thought I had understood the issues, but it was all on an intellectual level," he said. He stood there on the pad, the forest around him hanging on to some of its fall colors, and smelled the diesel fumes coming from ten compressors. He had a little trouble breathing. And just like his first visit, he began to extrapolate this industrial process across Gastem's forty thousand acres and then across hundreds of thousands of acres from New York's Finger Lakes down across the state's Southern Tier. "It was pretty staggering. It made me pretty sick," he said.

By this point, he said, the New York chapter's relationship with the national Sierra Club was "chilly." The state's executive committee a month earlier had voted to call on the state legislature to ban fracking. The San Francisco leadership called the local leaders and told them to stand down. It violated the one-club philosophy. The local chapter had to reverse its position. But the state chapter kept hearing from its members, many of whom were being visited by landmen seeking to lease their properties. A New York City councilman began raising the specter that upstate gas drilling could jeopardize the watershed that provided clean water to ten million people and require a $10 billion filtration plant. And Downs was educating himself about fracking. He began to worry about the benzene, a carcinogen, and other chemicals that were part of the fracking cocktail.

The issue came to a head in January 2010. Cornell University was holding a ceremony to inaugurate two natural gas–fired turbines to provide power and heat for the school, located three counties west of the Ross #1. Cornell had relied on coal to generate heat, but as part of an effort to reduce its carbon emissions, it switched fuels. Bruce Nilles, the head of the national Sierra Club's "Beyond Coal" campaign, was there to congratulate the school. So was Kate

Bartholomew, a high school biology teacher who was the volunteer chair of the Sierra Club's local group. She confronted Nilles about the Sierra Club's tolerance of natural gas and had what she later described as a tense moment. A month later, National Public Radio interviewed her about her opposition to fracking. "I don't want to walk out and see five-acre drill pads all over my hillside. Yeah, and I don't want my water to be contaminated," she said. The reporter also interviewed Carl Pope for the segment, who conceded there was some friction with local New York chapters.

He went on to make a statement that was a step too far for many Sierra Club members in New York. "What's happening with the new discoveries of natural gas is that parts of the country that historically didn't pay any environmental bill for energy production because they didn't produce energy are going to start paying a bigger share of the bill," he said. When Roger Downs heard this on the radio, he became upset. Carl Pope was betraying his members, he thought. "That is a pretty stark thing for an environmental leader to say, to think we have an obligation to despoil our own turf. It is a hard pill to swallow," he recalled. Pope seemed to be saying that to use more gas—and therefore less coal—people in New York and elsewhere had to suck it up and accept the wells and the pads. "It was a real turning point," Downs said. "Carl got a beating from that point on." Pope had already said he was stepping down as executive director but would remain as the club's chairman. By late 2011, the national Sierra Club's board of directors had decided he would need to leave this post as well and cut all ties with the organization that had been his home for eighteen years.

With Pope no longer at the helm, local antifrack activists began to dismantle the Sierra Club's endorsement of natural gas. At the time, other environmental groups were also backtracking. Prominent environmentalist Robert F. Kennedy Jr., president of

260 | THE BOOM

Waterkeeper Alliance and nephew of President John F. Kennedy, hailed the "revolution in natural gas production" in 2009 and said that it "has made it possible to eliminate most of our dependence on deadly, destructive coal practically overnight." Two years later, he changed his view, saying the gas industry and government regulators weren't doing enough to protect the environment and safeguard communities. The antifracking movement grew as events raised questions about the competence of the energy industry. In April 2010 the Deepwater Horizon drilling rig exploded in the Gulf of Mexico. For nearly three months, the industry failed to cap the steady torrent of oil flowing out of the damaged well. An undersea camera broadcast a live feed, which played around the clock in the corner of cable news channels. It was the first oil spill given a close-up in the era of reality television. The industry's futile attempts to shut down the well eroded the public's trust of energy companies and their promises that they could drill wells in an environmentally sound manner.

As this infamous well dominated cable news from late April until mid-July, a driller lost control of a gas well in Clearfield County, Pennsylvania, sending a geyser of gas, water, and chemicals into the air. A blowout preventer failed to contain the well, just as the blowout preventer had failed to shut down the Macondo well that the Deepwater Horizon was drilling. Three weeks later, in June 2010, Josh Fox's documentary *Gasland* premiered on television. It showed striking images of residents near gas wells lighting their faucets on fire, a claim later contested credibly by the energy industry and state officials. Fox emerged as a skilled provocateur, challenging the industry's statements that they were responsible operators, and a leading figure of the antifrack movement.

His documentary was like a match struck in a bone-dry forest. It aired amid daily doses of the Deepwater Horizon and BP's inept attempts to shut down the well. And distrust of the fossil-fuel industry grew with every report of the rapidly melting Arctic ice. Communities in upstate New York were split between people who wanted the economic benefit of drilling and those opposed who had moved there because it was quiet and bucolic. A *New York Times* article about the actor Mark Ruffalo captured the split. Ruffalo lives on a forty-seven-acre former dairy farm in upstate New York where the industry wanted to drill wells into the Marcellus Shale. Ruffalo led a crusade to keep out the rigs. His neighbor, dairy farmer William Graby, was quoted as describing him as an outsider with a big bank account. "We need industry and jobs so we can send our kids to college," he told the reporter.

While this conflict played out in the communities above the shale, opposition grew in cities and on college campuses. "The fracking issue has become a huge grassroots issue. You haven't seen anything like it since the antinuclear movement in late 1970s and early 1980s in terms of real, community oriented, grassroots activism," said Alan Nogee, who headed the Union of Concerned Scientists' clean energy and climate program for twelve years.

By the end of the summer of 2010, the antifracking movement had coalesced and gained strength. Embracing natural gas as a fuel that could help reduce coal consumption and enable the growth of renewable energy became difficult for environmental groups without losing their grassroots supporters. "National environmental groups really don't have a choice if they are going to remain viable," said Nogee. Only the Environmental Defense Fund, among the large national environmental groups, remained committed to natural gas and worked hard to get the industry to adopt better drilling practices.

In Albany, Downs grew increasingly leery of any potential benefits of natural gas. He worried that fracking allowed too much natural gas, itself a greenhouse gas, to escape into the atmosphere. He also wondered if natural gas was really a bridge fuel, connecting the coal-dominated present to a renewable energy future. The gas industry was doing such a good job of finding gas that it was creating a glut and driving down prices. Gas wasn't just pushing coal off the power grid. It was also making it harder for new wind and solar plants to compete. "The glut of cheap natural gas was directly competing with renewables. As much as we enjoyed the notion it was thrashing coal, it also provided a new barrier to wind and solar plants," he said.

Over time, Downs kept an eye on the Ross #1 well, an hour southwest of Albany, where his personal opposition to fracking had begun. He got to know local landowners and obtained numerous documents. Eventually nine residents claimed that their water wells had been contaminated and that some horses had died. But these claims bothered Downs. Gastem and the state had tested the water wells extensively before drilling had begun—and continued testing afterward. "There was not a whole lot of difference in terms of water quality," he said. The rural water wasn't pristine to begin with, and there were no signs that fracking chemicals had entered the nearby well water. "The battery of tests wasn't exhaustive," he said, "but you would assume some of the chemicals would turn up."

He felt certain the residents weren't fabricating their claims, and none ever brought a lawsuit or sought compensation. "There was something really genuine about the community distress and the loss of quality of life. The truck traffic, it is a very rural area, dirt roads. It would keep them sleepless. The noise, the smell was very traumatizing. And this was touted as a completely successful, environmentally safe experience. I have to say the testing confirmed that. Yet universally, the neighbors complained of health problems," he said. "I didn't know what to make of it. I didn't sense they were lying. It has been difficult for me to reconcile. I want empirical proof, yet I can see something really bad has happened to these folks. But I can't prove it. That's a really hard position for me with my science background."

With the appointment of Michael Brune as executive director, the Sierra Club's position on natural gas changed. He believes that support of fracking isn't a tenable position for the country's largest environmental group. "Fracking has exploded around countries, hundreds of thousands of new wells being drilled in regions or communities that haven't had that industrial production of energy. . . . People are confronted with the process of energy production in a way that they haven't in the past," he said. "There is great concern about the degradation of key ecological landscapes. There are also concerns about the impact of fracking on water supplies air quality and on greenhouse gases." By 2011, the Sierra Club had overhauled its energy policy and produced a document significantly more critical of natural gas. The new policy goal was to "develop and use as little natural gas as possible." The Sierra Club's new preference was to replace existing nuclear and coal plants with renewable energy, not natural gas.

Brune said this change is partly because of a rise in confidence about renewable energy. In 2007, when Carl Pope first accepted Aubrey McClendon's donations, electric power from wind, solar, and plants that burned biomass made up 2.7 percent of the US power supply. By the fall of 2012, when Brune made these comments about fracking, the figure had risen to 3.7 percent. Electricity from coal, which the Sierra Club viewed as the number one threat to climate change, had dropped from 48.9 percent to 38.7 percent. Natural gas had risen from 23.5 percent to 33.6 percent. Renewable energy was chipping away the dominant position long held by coal. Natural gas was smashing it with a sledgehammer.

I asked Brune if he considered natural gas an ally in the fight against climate change and carbon emissions. "If you were to have asked me does gas help or hinder our efforts to stabilize greenhouse gas emissions and reduce them, I would say yes, it both helps and hinders," he said. Low natural gas prices are helping speed up the retirement of about one hundred coal plants across the country. "If you think that gas isn't playing a role in the downsizing of the coal industry," he observed, "then you are not really paying attention to how the power sector is working. But at the same time, if you think that a large reliance on natural gas to displace coal or replace coal won't have an adverse effect on our climate and greenhouse gas emissions, then you also don't know much about how gas actually works, or how fracking actually works, and what the environmental costs of gas are likely to be."

Imagine you could snap your fingers, Brune said, and speed up the production of shale gas in a way that limits its impact on land, water, and communities. Global use of gas increases by 50 percent over the next couple decades and overtakes coal as the second-most used fuel sources (though still trailing oil). The International Energy Agency, a Paris-based organization that analyzes energy policy for the world's leading industrialized nations, explored that scenario. It still wouldn't be enough, Brune said, citing the IEA's conclusion. The amount of carbon dioxide in the atmosphere would still rise to 650 parts per million, from 392 parts per million, the level at the time. That "would lock us into 3.5 degrees Celsius of warming, which is catastrophic for the planet," he said. The only viable option, he continues, is to use wind and solar to meet any new energy demand.

Environmental activist Bill McKibben has an even starker view. Using gas to displace coal might have made a difference twenty-five years ago, he said. "We are so close to the edge now that what we require is the very, very quick conversion not to somewhat cleaner fossil fuels but off fossil fuels altogether," he said. According to McKibben, natural gas is a half measure, and the crisis is so severe that we can't waste time with half measures anymore.

Pope sees the situation differently. Natural gas, he said, has made a significant dent in carbon emissions. He calls the collapse of coal "a fantastic success story."

Back in January 2007, before natural gas production began to surge and just a month before Aubrey McClendon launched his "Coal Is Filthy" ad campaign, US energy consumption released 543 million metric tons of carbon dioxide into the atmosphere in a single month. That includes coal power and natural gas power plants as well as what came out of the exhausts of airplanes, cars, and trucks. Five years later, in January 2012, the United States sent 479 million metric tons into the air. That's a 12 percent reduction. If you drill down into the data, you will discover that coal plants released 48 million metric tons less, a 25 percent reduction. Better fuel efficiency from cars running on gasoline and diesel led to a 17-million-metric-ton drop, down 11 percent. Increased use of natural gas to generate electricity, run factories, and heat homes led to a 14-million-metric-ton increase, or about 11 percent.

The United States is one of the few places in the world that decreased its carbon output—and decreased it dramatically. It never ratified the Kyoto Protocol—a United Nations effort to enact greenhouse gas emission reductions—but is on pace to exceed the targets anyway, even though many signatory nations have failed to meet their obligations. The International Energy Agency, forecasting ahead a couple decades to a world with more abundant gas, believes the United States and other major industrialized nations will lower their CO_2 output, although increased energy consumption in China and the rest of the world offset that gain. An increased use of natural gas, by itself, "cannot on its own provide the answer to the challenge of climate change." That's the point Michael Brune and Bill McKibben both make. But the IEA looked at an alternative scenario, with less gas and more coal. The result, predictably, was higher carbon dioxide emissions. And more could be done to lower natural gas's greenhouse footprint: use pumps and compressors with better emissions controls and eliminate gas venting, a practice the industry uses to drill wells quickly. People worried about climate change "may find it difficult to accept," the IEA report stated, but burning more gas helps limit emissions.

Some argue that natural gas can help reduce carbon emissions. Others counter that it's not enough—and the reduction isn't nearly quick enough. They argue for a large-scale switch to wind, solar, and nuclear, none of which emits carbon. The two positions may not be mutually exclusive. The cost of renewable energy is falling quickly and is increasingly the same price as coal, gas, and other conventional fuels—even without subsidies. This will mean more demand for renewable energy. But since most renewable power comes and goes with clouds and the wind, an energy analyst for Citigroup noted in a large report to investors, more renewables will, in turn, drive demand for more gas-fired plants that can turn on and off quickly. The analyst said that gas and renewables formed a "symbiotic relationship."

John Hanger has spent a lot of time thinking about the future of energy. He spent his career as a lawyer, advocate, and government official, all the time focused on energy. He represented Philadelphia's poor when the local gas company wanted to disconnect them for failure to pay their bills. He founded Penn Future, a Pennsylvania environmental group. He served on the state's utility commission and championed electric-power deregulation. And he was the state's environmental protection commissioner when the Marcellus Shale drilling boom began in earnest. He's also an inveterate blogger on fracking and energy topics, often punctuating interesting facts with "Wow!" or "Amazing!" In late 2012 he announced a long-shot candidacy to become governor of Pennsylvania in 2014.

A year after Aubrey McClendon riled up a crowd of energy executives at the Pennsylvania Convention Center while protesters outside chanted "Hydrofracking's got to go!" I attended the same conference. Again there were protesters outside with signs and speeches. Inside, an Exxon executive placed the development of shale gas in the pantheon of great American innovations, such as the lightbulb and the personal computer. The two sides' positions had hardened. Communication and compromise were elusive. John Hanger and I left this deadlock and walked a couple blocks away to get lunch. He started off by describing himself as a pragmatist. He thinks environmentalists who don't want to push natural gas to eliminate coal are "tragically mistaken." Spearing some pasta on a fork, he said, "I love renewables, but it is just irresponsible not to capture the environmental benefits of gas today. It is real. It is huge."

People might not want to hear this, he went on, "but Aubrey McClendon may have done more to cut climate emissions—while saying he doesn't believe it's a serious issue—than most environmentalists, individually or cumulatively."

Natural gas is here today. It's a powerful, nimble fuel that can sate modern society's desire for electricity to run smart phones, heat homes in the winter, and provide base materials for an unending supply of plastic goods, from car bumpers to toys, and even to parts of the microphone that Josh Fox used, for a second year, to address the crowd on the streets outside the conference. But even the abundance brought about by fracking the source rocks won't last forever. Proponents like to talk about a hundred-year supply of natural gas in the United States. Few people who have spent time studying gas markets take this seriously. I don't believe it. The new gas is real, but so is rising demand from new manufacturing facilities, power plants, and exporters looking to send it to Asia. At some point in my lifetime—and certainly in my children's—natural gas will become more scarce and expensive. The only question is what comes next.

12

GHOST RIDIN' GRANDPA

In July 2007 Brian Smiley uploaded a two-minute video of his grandparents to YouTube. As it begins, his seventy-seven-year-old grandfather, dressed in a red, white, and blue shirt with stars and stripes, sits in the passenger seat of a car. He turns to his wife and asks, "Well, Grandma, what's on our schedule today?"

"Well, it's pretty much open," she responds. "What do you think we should do?"

Jangly country guitar music plays in the background.

"I'm fresh out of ideas. You don't have any ideas?" he asks, looking bored.

"Well, there's one thing we could do," she says. The music ends with the sound of the needle being ripped off vinyl. She turns to her husband, her mouth drops open in excitement, and says, "We could ghost ride the whip."

A rap song begins with a thumping, distorted bass line. Smiley's grandparents dance in the street on both sides of the canary yellow car as it rolls driverless down a street bordered by towering pine trees. The grandfather wears sunglasses and some sort of a doo-rag on his head. For the next minute, the couple dances as the car continues to roll forward. He puts his hands on his knees for an improbable few seconds of the Charleston. She grabs the hood and kicks her feet out. It is a suburban send-up of a dangerous Bay Area tradition.

The clip went viral. It had one million views in a year and another million the next year. I first saw it when a Houston energy investment bank included a link to it in an email to clients and reporters. The bank noted that the man was none other than Dr. Claude E. Cooke Jr., a "legend of hydraulic fracturing." I met Cooke a couple years later, after he'd called me to talk about fracking. After a couple meetings, in which he wore immaculately pressed Oxford shirts, it occurred to me that he was the grandfather in the "Ghost Ridin' Grandma" YouTube clip.

I watched the video again, amused by this strange connection. His claim that he was fresh out of ideas couldn't be further from reality. Cooke has two dozen patents. Even in his ninth decade, he is filing new ones. This man, who had achieved a level of Internet fame alongside cute piano-playing cats and teenage skateboarders with questionable judgment, is an industry legend. In 2006 the Society of Petroleum Engineers honored nine men (they were all men) as "Legends of Hydraulic Fracturing" for seminal innovations. While most of the honorees had a single contribution, Cooke had three.

When Cooke first called me, he wanted to talk about how to build a better well. Fear that fracking itself could contaminate water, he argued, was misplaced. The cracking of shale generally takes place a mile or two underground and thousands of feet below freshwater aquifers. Getting to that rock, however, means drilling a long hole in the ground. A slow migration up through rock strata that would take thousands or millions of years can occur in minutes through a well. In its search for hydrocarbons, the industry builds superhighways that traverse geologic epochs. Worry about the wells, he said, not fracking.

A couple weeks after our first phone conversation, we met in his office in Conroe, Texas, a few stops on the interstate north of downtown Houston and near George Mitchell's Woodlands development. On the walls are black-and-white photographs of a 1970s frack job in the Texas panhandle. The photos show Halliburton and Exxon engineers studying a pressure gauge that looks like an old-fashioned Hollywood movie camera. A young Cooke stands in the middle of the pack of engineers. "If there is a problem, the issue is well integrity," he explained, fixing me with a hard stare through wire-frame glasses. If something goes wrong with a fracked well, the likely problem is faulty cement. Cooke said that people concerned about fracking were trying to fix a problem that didn't exist and ignoring the problem of poorly built wells that was staring them right in the face. If the industry built better wells, he said, you could eliminate problems with water contamination.

The purpose of a well is to reach deep into the earth and suck out long-buried oil and gas. But wells must also be constructed to prevent unintended movement of salty water or contaminants. The industry takes many steps to make sure that the open hole it has drilled is contained, constricted, checked, and controlled. It inserts steel pipe, called casing, into the open holes and locks it into place. Then it pumps in cement to secure the pipes and achieve "zonal isolation." This is industry jargon, he said, for keeping the gas, or salty water, in one rock from flowing into another. The cement may sound mundane, but it is big business in the oil fields. The energy industry spends about $105 billion annually to extract energy from North America using hydraulic fracturing. About $5 billion of the $105 billion outlay is spent on cementing. This is not off-the-shelf cement found at Home Depot. The oil industry uses an extensive selection of densities, additives, and ingredients. A standard guide to oil-field cementing runs 171 pages.

How many wells have been constructed to withstand a lifetime in the earth? How many have effectively created a single pathway for oil and gas to flow up the inside the casing? And how many have left channels and holes in the cement outside the pipes? Considering how important these questions are, there are no satisfying answers. There is a subsubspecialty of oil-field services called, prosaically, cement evaluation. Only a tiny fraction of the $5 billion spent on cementing is used to evaluate the cement itself.

There are engineers and executives in the industry who share Cooke's view of the importance of building wells right. So does Mark Zoback, a Stanford University geophysicist who served on the US Energy Department committee that studied shale production. To eliminate risks, he told me, "There are three keys—and those are well construction, well construction, and well construction." Industry studies and experts concur that cement in wells fails regularly to one degree or another, although rarely catastrophically. The failures tend to be small and subtle but significant enough over time to cause problems. Gas seeping through faulty cement can get into shallow aquifers, infusing them with high levels of methane. Residential water wells that tap these aquifers can pump flammable gas up into homes. Usually, regulators will shut down the well and require the home to use water from refillable plastic "buffaloes" that sit outside the home. Sometimes, expensive ventilation pipes and filtration systems can restore potable water. In other cases, the well is lost, rendering the home uninhabitable. This is what happened in 2010 in Bradford County, Pennsylvania. State investigators found "improper" cementing and well construction. Chesapeake, the driller, paid a record fine to the state. Later, after a lawsuit, it bought out the homes of sixteen families, who moved elsewhere.

Toward the end of my first meeting with Cooke, I asked him what he would demand if he owned land and leased it to be drilled. "Well, I would be pretty sticky about it," he responded. He said he would demand proof that neither gas nor liquids were flowing through the cement that encircled the steel pipe. I asked how he would do that. He stood up and walked to his door and politely asked his assistant to bring a two-page printout of his patents and publications. She came in and gave him the document, which he passed to me. "You see number ten?" he asked. I looked at the printout and saw a reference to a 1979 paper from the *Journal of Petroleum Technology*. It described a "new tool for detecting and treating flow" in the cement. "I would tell them, 'I want you to run that tool in the well,'" he said.

There was only one problem. The tool—his creation—wasn't available anymore. It had disappeared from the market. Cooke then told me he was thinking of trying to bring it back.

In 2010 Cooke woke up in the middle of the night thinking about cement and his tool for the first time in a quarter century. The Deepwater Horizon was in the news. The giant offshore floating platform had lost control of the well it was drilling in five thousand feet of water, nearly fifty miles from the Louisiana coast. The resulting explosion had killed eleven workers. BP scrambled for months to cap the half-finished well sitting on the seafloor. Millions of barrels' worth of oil flowed into the Gulf of Mexico, making it "the worst environmental disaster America has ever faced," according to a presidential commission that examined the disaster.

I covered the Deepwater Horizon for the *Wall Street Journal*. Early on, much media attention focused on the failure of the blow-out preventer, a several-story-tall set of valves on the ocean floor that was supposed to deploy giant shears to cut and seal off the well. It came within two inches of clamping shut, but it didn't close off the oil flow. Soon after the blowout, many petroleum engineers began to suspect the root problem was that something had gone wrong with the well construction, and, in particular, the cement in the wellbore. Cooke was among the industry insiders who zeroed in on the cement. As the well gushed crude into the gulf like an open wound, Cooke woke up thinking about the well. He thought about his invention, called a radial differential temperature log, or RDT, and whether it could have prevented the disaster. He decided the answer was likely yes.

It is impossible to know if his tool would have detected cement problems on the Deepwater Horizon, but it is possible that using better cement evaluation tools would have helped. A couple days before BP lost control of the well, the company's foreman on the rig ordered a full suite of diagnostic tools to make sure the cement had set and formed a solid seal. Oil-field service company Schlumberger sent a team to the rig by helicopter. To most people, Schlumberger is not as well known as Halliburton, which vaulted into the public eye after Dick Cheney, its chief executive, left the company to join George W. Bush's presidential ticket. Despite a lower profile, Schlumberger, or Big Blue, as it is known, is the largest Western oil-field service company in the world, followed by Halliburton, aka Big Red. Schlumberger's crew arrived on the Deepwater Horizon on April 18, 2010, two days before the well blowout. BP had ordered a cement bond log as well as more sophisticated tools, including an isolation scanner, which is similar in some respects to Cooke's tool.

The crew members waited for two days to run the tests and then BP sent them home. They departed by helicopter ten hours before the explosion. The tests were never run. Schlumberger charged BP $128,000 for its workers and tools, even though they didn't do anything. The bill was a rounding error for a well that cost about $100 million. However, the well was running over budget and had taken more time than expected. Running the Schlumberger tests would have required eight hours. BP relied, instead, on another test to determine if the well was secure. The test results were confusing and anomalous. Instead of stopping work to figure out what was going on, workers aboard the drilling rig decided to press ahead. The federal government would later file criminal charges against two BP workers responsible for running and interpreting this test. The cement had not created an effective barrier, and without detecting this failure, BP and the crew were dangerously vulnerable.

Shales aren't high-pressure reservoirs, like the one encountered by the Deepwater Horizon drilling rig. When BP's drill bit reached the targeted sandy reservoir, a combustible mixture of natural gas and petroleum liquids in the pressurized "pay zone" pushed its way up and out of the well and onto the floor of the drilling rig, where it ignited with lethal results. Drill into shale and nothing will happen. Companies need to smash the shale into submission before it gives up its hydrocarbons. Cooke understood this distinction. But the basic principle of well integrity is the same, he thought. If a company spends the time and uses tools to determine if a well is cemented, the well will be safer for the workers and have a much lower chance of contaminants coursing through tiny channels outside of the pipe, reaching the surface, or finding a way into a shallow drinking-water aquifer. "A channel doesn't have to be very big to carry a lot of fluid; a finger is enough," said Cooke.

The problem with a tool like Schlumberger's isolation scanner is that it is so expensive. It is generally used only in the industry's most challenging wells, if then. It relies on four transducers that emit high-energy pulses. The tool Cooke invented was simpler. It was a fancy thermometer. And as Cooke told me once by email, "Temperature measurements are cheap." Because the tool is less expensive to manufacture, Cooke believes it could be widely deployed in the thousands of shale wells being drilled every year. Each could provide a measure of assurance that the well wouldn't leak and leave behind an environmental mess.

The shortcomings with cement—both in deepwater wells and less complex onshore wells—are one of the industry's best kept secrets. The industry talks regularly about the protection against dangerous blowouts and groundwater contamination that cementing wells provides. But cementing, despite huge technological improvements, remains an imperfect science. In an exhaustive report of the causes of the BP Deepwater Horizon offshore catastrophe, a national commission noted that "cementing an oil well is an inherently uncertain process. . . . Even following best practices, a cement crew can never be certain how a cement job at the bottom of the well is proceeding as it is pumped. Cement does its work literally miles away from the rig floor, and the crew has no direct way to see where it is, whether it is contaminated, or whether it has sealed off the well." While it sets, the cement is exposed to extreme heat, pressure, and contaminants. Hairline seams can appear in the cement, or even larger finger-sized holes, that undermine cement effectiveness. The cementing crew is left with incomplete and indirect measurements. It can be like trying to determine someone's gender by looking at his or her shadow through a telescope.

"Why doesn't the oil industry pay more attention to cementing problems?" Cooke asked rhetorically during one of our meetings. "I answered my own question over a period of time: because it costs money to do it, and there is no pressure to do anything."

Cooke was born in El Dorado, Arkansas, in the midst of an oil boom. Nearly everyone's father worked for an oil company. His dad worked on wells for Magnolia Petroleum, a forerunner of today's Exxon Mobil. Pride in the dangerous, tiring work passed from one generation to the next. Cooke recalled that in the small oilfield elementary school, if a kid made a disparaging remark about your father's employer, there would likely be a fight. While his own father's formal education stopped short of a high school diploma, Cooke's mother had higher aspirations, first for herself and then for her son. She attended college and wanted a degree in chemistry, but her father didn't believe that science was an acceptable profession for women. Settling for a teacher's certificate, she poured her dreams into her son. From the time he was a little kid, Cooke said his mother told him, "You are going to get a PhD. You are going to have plenty of education."

Cooke never strayed from the path his mother laid out for him. After attending Louisiana Tech University, he went to the University of Texas and earned his PhD in a field of physics that dealt with the interaction of molecules. Then he returned to the industry that dominated his youth. For three decades, he was a standout at Exxon's research facility in Houston, a concentration of talented scientists who threw off innovations that helped transform a superstition-soaked industry from one that ran on hunches into one that embraced science to solve problems. When Cooke started work at the research center, in 1954, there were still wildcatters around who used witching twigs to find oil. By the time he left in 1986, the industry could send sound waves through thousands of feet of rock, process the bounced-back signals through some of the world's fastest computers, and find oil. This technique was invented at Exxon's research facility in Houston. "I spent thirty-two years there, and I enjoyed every day of it," said Cooke.

Over the years, he made his share of innovations. One was the radial differential temperature logging tool. It looks like a long, skinny metallic pool cue that drillers lower into oil wells to figure out if the cement hidden away behind the pipes had set properly. It is a high-tech whirligig on the end of a long steel tube that spins in the well and can sense minute fluctuations in temperature that are telltale signs of water or gas flowing through cement. In most wells today, companies run what's called a cement bond log. The CBL uses acoustic signals to "listen" to the pipes. Cement that has adhered to the outside of the pipe, creating a solid seal, makes a different sound from cement that has left even a microscopic gap. But a cement bond log can't tell if there is a channel an inch or two away from the outside of the pipe where high-pressure gas is flowing. The cement bond log, said Cooke, gives drillers a false sense of confidence.

"What the industry says will determine whether or not a cement job is adequate, I don't agree with. The industry says you run a cement bond log," he said, banging his open palm on the table to punctuate the last three words. "They say if you get a good bond, you get a good cement job. That is not true. Not true." Another two palm slaps. "You can have flow in the annulus, in channels, through the cement, or between the cement and the formation." The cement bond log is a good test, he said. "It is necessary, but not sufficient as a mathematician would say." The industry is fooling itself—and fooling the public, he believed. The industry needs to run better cement evaluation tools down wells. Think of it as preventative medicine, he said, to find problems before they spin out of control.

The government issued Exxon Patent 4,074,756 in 1977 for his invention. A series of field tests showed that it was better than a CBL at finding leaks. Cooke was canny. He didn't pitch the tool to his Exxon bosses as a leak-detection device. Instead, the tool's original purpose was to determine if the cement had created a good enough seal to keep oil production healthy. Not long after he had a working prototype, an Exxon colleague working on the sprawling King Ranch in South Texas called him. "There's too much water in the well, and it's choking off the oil flow," his colleague said. "Get down here with that tool and figure out where the water is leaking into the well." Another call came from an Exxon team in Germany. For a year, he went around the world with the tool, running it down Exxon wells. As the RDT was lowered into the well, the tool deployed two small prongs that pressed up against the inside of the well and then rotated in circles. This rotation gave a 360-degree view of what was hidden behind the pipe. Since a small channel of gas or water will create a slight fluctuation in the pipe's temperature, he calibrated the tool to detect a difference of less than 0.01 degree Fahrenheit. "In many cases, I found flow behind casing, in some places where it had not been known. There was no other technique that would have detected it," he recalled. With his RDT, he could accurately find leaks and shut them down by puncturing the well and squeezing in new cement.

After he demonstrated its usefulness and wrote up the results in a couple papers published by petroleum engineering groups, Cooke found a manufacturer. Within a few years, there were thousands of RDTs deployed all over the world. Then in 1988 the manufacturer fell on hard financial times and was acquired by Halliburton, a larger competitor. Halliburton decided to stop making the tool.

By that time, Cooke had moved on to other inventions. He developed and patented a ceramic bead that was stronger than sand and useful for fracking deep wells. This invention earned millions of dollars for Exxon. He also did important work on using vibrations to improve cement quality. In the 1980s he performed pioneering studies on how oil-well cement sets. Twenty years later, the American Petroleum Institute, the industry's lobbying powerhouse and the final word on drilling and building wells, issued new guidelines on how to cement wells. The document praised Cooke's work, calling it a "revelation" and "one of the industry's most important publications for the advancement of cementing technology."

He left Exxon after earning a law degree at night and began his second career as a patent attorney. He worked for Baker Botts, a white-shoe law firm, for a decade and then cofounded another energy-focused law firm. In 2011 he left behind the long hours of supervising dozens of lawyers and set up a small patent law practice in Conroe. He remains a wiry and energetic man. Despite two decades practicing law, he remains at heart a scientist. "It's like your first girlfriend. You never quite get over her," he said. "I was trained as a scientist. I was looking at the science of wells. What is the data regarding sealing of wells. All the data that I have says cement is not a reliable seal." If the scientist in him realized there was a problem, it was more a sociological bent that got him wondering why this problem hadn't been fixed—and why his radial differential temperature tool, or one similar to it, wasn't a staple in the industry's toolbox.

A few months after he awoke at night thinking about cement, Cooke made a decision. He would reintroduce his tool to the industry. The first step was to build a modern prototype of the tool, so that he could demonstrate how simple and effective it was. Surely, he thought, if he could show the industry that it could prove its wells were solidly built, their self-interest would take over. The gas industry faced mounting public relations headaches. If the backlash grew any stronger, Cooke figured, the industry's new profit center could be stopped. It was a matter of self-preservation.

Cooke knew just the man for the job of building a newer, better RDT: the same man whose company had made the original in the 1980s.

M arvin, are you doing all right?" asks Cooke, extending his hand to Marvin Gearhart.

"I think so, but how do you tell?" Gearhart responds.

The men stand in the foyer of Gearhart Companies, in an industrial park south of Fort Worth. A cardboard cutout of John Wayne looks on. "He's our security," Gearhart jokes in a pronounced Texas drawl. Gearhart waves Cooke back into a conference room, and they sit down at a large wooden table.

Gearhart is every bit as much a legend as Cooke. Two years after the Society of Petroleum Engineers named Cooke as one of the pioneers of fracking, it honored Gearhart and four other pioneers in drilling wells. Between them, Gearhart and Cooke have about a century's worth of experience in the oil field. I've tagged along for their meeting. I'm not convinced the RDT tool—or a modern version—is the answer to building better wells, or that these two octogenarians can pull it off, but I want to hear what Gearhart has to say. His life's work has been building tools for the oil industry.

There's an easy rapport between the two men and many similarities. Like Cooke, Gearhart had grown up in the oil fields. His father worked drilling wells in southeastern Kansas. Cooke and Gearhart were born two years apart and both recall hitchhiking to high school in the days before yellow school buses became common. Both went to work in the oil industry right out of school. Physically, they are different. Cooke is tall and so skinny that he regularly grabs his belt while standing to make sure his slacks aren't slipping. He stands up straight and erect. Gearhart is shorter and stouter and peers from beneath large, white, bushy eyebrows. He hunches. Cooke spent most of his career working for large corporations and law firms; Gearhart was an entrepreneur. His first job after graduating from Kansas State University was working for a well-service company. His first assignment outside the office was to help Stanolind Oil on one of the world's first frack jobs. He didn't like working for someone else, so within a couple years, he borrowed some money and outfitted a truck with oil-field equipment, hired a couple employees, and created Gearhart Industries. Drilling came naturally to him. As a boy, he built a small rig to drill water wells for neighbors. "I got it in my blood," he says.

By the time the men first met in the 1970s, Gearhart Industries was a major oil-field service company. At its height, it controlled about 16 percent of the global wire-line business, manufacturing various tools that were lowered into wells by a cord. These tools did everything from testing rock pressures to evaluating cement. Gearhart Industries' wire-line trucks—mobile units with everything needed to probe and test a well—were in demand worldwide. The company grew quickly, especially after the 1973 OPEC oil embargo encouraged more drilling in the United States. In 1974 Gearhart Industries made $27 million in revenue. Within a decade, the company recorded $344 million in revenue.

Then came the 1980s oil price collapse. Large portions of the industry went on life support. Gearhart went from cutting deals with General Electric to laying off employees and scrounging up work. Gearhart Industries' stock price fell. Predators began to circle. Smith

International, a competitor in the oil-field service industry, started buying up Gearhart shares on the way to a hostile takeover. To ward off Smith, Gearhart purchased a smaller company that used sound waves to search for oil. When business didn't pick up, the debt used to buy the company became unmanageable. Hobbled and nearing bankruptcy, Gearhart agreed to be purchased by Halliburton in 1988 for $277 million and the assumption of its debt. Gearhart started over, building up another company that made drilling bits. He sold this company in 2005 and started Gearhart Companies, his third company.

Cooke and Gearhart spend a few minutes catching up, and then talk turns to building a new version of the radial differential temperature tool with modern electronics. "I wanted to bring you one of the old tools," Cooke says, so Gearhart's engineers can take it apart and work on a new prototype. He thought he had one in a storage space above his garage at home, but he had looked, as had his grandson, and it couldn't be found amid the family Christmas decorations. He was planning to send up a granddaughter, he says, because women can find anything. The next few months also failed to turn up anything. Finally, after his meeting with Gearhart, he located one in Bakersfield, California. He called the owner of the company there, Well Analysis Corporation, looking for someone able to manufacture the tool. The owner said he remembered the tool and had one in his shed out back. A few days later, an eight-foot-tall box arrived in Texas. The tool had broken after years of use, and Well Analysis had cannibalized some parts from it to repair other tools. But the key instrument, the rotating sensor, was still there.

Cooke asks Gearhart why the tool disappeared from production. "Halliburton bought you, and the tool disappeared," says Cooke. Gearhart nods in agreement. "What I've always wondered about is why no one else picked it up," Cooke remarks.

"It probably wasn't profitable enough. Companies like to push tools that are more profitable," says Gearhart.

Cooke came to the meeting with two grandsons. One, dressed in shorts and sneakers, is as quiet and unobtrusive during the meeting as a six-and-a-half-foot man can be. "My driver," Cooke explains. The other grandson is the one who instigated the YouTube video and told a contact at the energy investment bank about it. Brian Smiley, a business school graduate in his twenties, quit his job at the Boston Consulting Group to help his grandfather's quest. Together they started a business, the Well Integrity Technology Company, to put the tool back on the market.

Smiley pipes up to steer the conversation back to cement evaluation tools. One of his roles at the new company is market research, something he did as a consultant. He raises the question again of why the tool had disappeared from the market. If he can understand what happened, he figures, this will help him understand how to position it now. Perhaps, he suggests, bad cementing "wasn't a problem that companies wanted to know about?" Smiley tells me later that he wonders if there would be resistance to the tool because it is too good at finding problems in the well—problems that need to be fixed.

"Industry is not asking for this tool," says Cooke. Gearhart mumbles his agreement. Cooke continues: "That's our basic problem. If we run a tool that finds a channel, we have to do something. There's an obligation to do something about it." And doing something—trying to squeeze in more cement to fill in holes—can get expensive.

Gearhart suggests that running the tool down a well to look for leaks is "like buying insurance." This idea perks up Cooke. "The analogy I've thought of is it's like a vaccine," he says in an animated voice. The radial differential temperature log can find problems with wells before bigger problems crop up. "Not a large number of wells get this problem," he continues. "This is about the health of this industry. It's like giving the wells a vaccine, so problems are much less likely to occur."

Gearhart nods. But he is not willing to commit to working with Cooke on manufacturing a modern version of the tool. His current business makes a specialized tool. He's out of the general oil-field manufacturing business. And he makes clear to Cooke that while he is willing to help brainstorm, there will be no second go-around for the Cooke-Gearhart collaboration. He suggests that maybe a neighbor in the industrial park can help out. He picks up the phone and calls Dave Clark. A few minutes later, Clark arrives. He's in his fifties, with a salt-and-pepper mustache and reading glasses perched on his head. He works for Probe, which designs and manufactures specialized oil-field tools. Clark remembered the RDT. "At the time, it looked like it was a real good tool," he says.

Clark offers some off-the-cuff suggestions about how to make a new, modernized version of the tool. The electronics are smaller and better today, he says. Cooke grows excited and talks about all the different ways the tool could be useful. Clark leans back, his legs crossed and brow furrowed. Drilling is moving so quickly, he says. Companies want to drill a well and complete it as quickly as possible to get the oil and gas—and the money flowing. You're asking them to slow down and run tests that aren't required, he says, tests that could lead to delays. That's not how companies think, he says. "Here's how they operate: 'At this point, we don't care. If there's a problem, we'll fix it later.' It's like, 'Worry about that later on. I've got to get production going right now.' These guys are running their legs off."

Gearhart steers the conversation in another direction. "My hobby is magic. We need a little entertainment," he says, pulling out a deck of playing cards. On the back is a picture of John Wayne.

Gearhart shuffles them and hands them to me. Following his instructions, I create four piles and then take cards from each pile, distributing them to other piles. When I'm done, he tells me to turn over the top cards from each pile. I turn over four aces. Gearhart beams.

In the months that followed the meeting, as Cooke and Smiley continued work on bringing back the cement-leak detection tool, I researched how drillers approach well integrity and cement evaluation in shale wells. It's not a subject discussed widely within the industry. But I found a wealth of information about how a company had struggled with a leaky Marcellus Shale well drilled a year after the Deepwater Horizon blowout.

Flatirons Development drilled the well in May 2011 in Jefferson County, Pennsylvania. At first there were no problems. The well extended down for 6,700 feet and then for another mile horizontally through the shale. Flatirons, a privately held Denver company, wasn't trying to build an inexpensive well. It used three pipes, one set inside the other, following what is considered best industry practice. After each pipe was put in place, cement was pumped down through the pipe and back up the exterior, or the annulus, and given ample time to set and harden.

After drilling the well, the rig was moved to another location, and Flatirons workers left the site while a nearby impoundment was filled with water to frack the well. Frank Uhl, a retiree and board member of the local water company, went by the newly cleared pad and noticed a steady stream of gas bubbling up in a water-filled cavity around the well, like the fizzy surface of an open bottle of Coca-Cola. Gas was leaking up around the casing, stealing through what was supposed to be impermeable cement. Uhl thought of the underground aquifer that supplied water to a couple thousand people in the area. If the community lost its water, the nearest alternative was a hundred miles away.

Tensions between the Brockway Borough Municipal Authority and Flatirons soured soon after the company showed up in the Western Pennsylvania town. They didn't get much better when Flatirons, in February 2011, drilled into the aquifer and caused a nearby water well to go dry for a little more than a day. The water utility, after signing a legal agreement to allow Flatirons to drill on its property, went to court to stop the drilling, ending up with a brokered peace and various concessions from Flatirons. Locals were still unhappy with the Marcellus well. They felt bullied by the company. Someone expressed frustration by using some Flatirons heavy equipment left at the well site as target practice, the company said later.

It was a cool late-spring day as Uhl fumbled for his cell phone and called the water company offices to report what he saw. The news was relayed to Pennsylvania's oil and gas regulators. A state inspector arrived the next morning. When he got there, Flatirons was already back at the site, vacuuming up the water in the cavity. The inspector pressed Flatirons for a rapid solution. Due to the "possible exposure of the fresh water aquifer to methane contamination, it was necessary to quickly determine the condition of the cement behind the 5½" casing," Flatirons engineers wrote later.

Over the next couple months, Flatirons voluntarily did something unusual. It spent liberally to figure out what had gone wrong with the well—and left behind a valuable, detailed record. Flatirons hired the Schlumberger oil-field service company to run an ultrasonic image log, a tool that uses sound pulses well above the range of the human ear to find gaps between the pipes and cement. The test results were not encouraging. At points, the "cement was so spotty and unbonded," there could be multiple places where gas was flowing into the man-made annulus, according to Flatirons. "We saw gas-filled channels in the cement, but you can't really tell how large they are," Jeff Jones, a managing director at Flatirons, told me in an interview. Like blowing air through a straw into a thick milkshake, the gas had likely entered into the cement slurry before it hardened and created long channels that extended for hundreds or maybe thousands of feet. Flatirons collected a gas sample at the surface and sent it off for analysis, looking for isotopic fingerprints to determine if it had come from the Marcellus or from a shallower zone. The analysis showed it was coming from gassy sands only a couple thousand feet deep.

Flatirons then ran a noise log. Developed by Exxon in the 1970s at the same time that Cooke developed the RDT, a noise log is essentially a powerful microphone that listens to the well. It is so sensitive that it can hear a leak and help determine if the leak is a liquid, gas, or both. Flatirons worried that a lot of gas was moving up the well, possibly into the aquifer, and not making it to the surface. "The noise log is one of those things that you'd kind of rather not know, but we decided we needed to know," explained Jones. "If there was any way we were possibly moving gas into the aquifer, we needed to do something." But since the leaks were small, the gas couldn't be heard. "Most of the tools aren't set up to find a small quantity of gas," he explained.

The company also sent a sample of the cement to an independent lab for testing. The results weren't good. To ensure that cement works as advertised, the industry has established quality guidelines. One requirement is no more than twenty teaspoons of fluid be lost into the rocks from a set amount of cement in thirty minutes. If too much liquid escapes into the rock, the cement can become too dense to pump and won't go where it is needed. In the laboratory testing, the cement failed spectacularly. It lost over a hundred teaspoons. Flatirons determined that the cement had grown too chunky to spread evenly and provide a good barrier. Instead of a solid seal, it had left behind channels. But there was some good news. As these tests were run, the volume of gas leaking to the surface decreased, and testing the water aquifer didn't turn up any evidence of contamination.

Flatirons's investigation left me unsettled. The company used ultrasonic logs, noise logs, and even gamma ray logs to pinpoint the leak, but had failed to figure out where the gas was entering the wellbore. These tests cost nearly $200,000, according to Flatirons, and weeks of detailed work. If Flatirons had found a leak, it could have used perforating guns to blast small holes in the pipe and squeeze in more cement. Later that summer, in a presentation to the state, Flatirons argued against this remedy. It might seal the leaks, or maybe not, but it could also make it harder to frack the well. And it would cut into its profits. The well might have to be scrapped. The state decided that since the leak appeared to be going away on its own, Flatirons wouldn't have to try to fix the well.

Flatirons isn't the only company in the Marcellus to have problems with leaking gas. "Many other operators in Pennsylvania have been confronted with these problems," Flatirons engineers said in a paper they wrote about the well. The state convened a group of regulators and companies to study the problem. One issue the group has taken up is what is an acceptable level of leakage. The Flatirons well was leaking 270 cubic feet daily. The average US home uses 200 cubic feet daily. Flatirons pointed out state regulations for underground storage caverns that store gas for peak wintertime usage were allowed to leak up to 5,000 cubic feet a day of gas without any repercussions. The working group decided this was a good starting point. Leaks happen. Fixing them is hard.

There was one more tool that Flatirons could have tried: Schlumberger's isolation scanner. This is the superdeluxe tool that BP had ordered, and then skipped, aboard the Deepwater Horizon. Schlumberger introduced this tool in 2006 to improve on older tools that it says suffered from limitations and generated ambiguous results. Flatirons asked Schlumberger about using an isolation scanner, said Jones, the Flatiron executive, but was told none were available in Pennsylvania. To bring one in would quadruple the cost of running it down the well.

In Pennsylvania, the drilling industry doesn't want to be forced to use this tool. In January 2011 the state issued a draft of new rules on testing a well's mechanical integrity. The state reserved the right to require the driller use an isolation scanner. The Marcellus Shale Coalition, an industry group that lobbies on behalf of the state's drillers, requested the draft language be changed. Specifically, it wanted to replace "isolation scanner" with "ultrasonic logging tools." A month later, these less precise and less expensive ultrasonic logging tools proved inadequate in finding the Flatirons's well leak.

"If I had a more precise tool, maybe it would have showed us where there was a minor channel," said Jones. This might have changed his mind about ordering remedial cement. "Maybe I would have a different opinion if I had other tools available," he added.

The well was fracked, after a six-month delay, and the amount of gas leaking to the surface continued to decline. It is too small an amount to even flare off. "It is now down to the amount of gas that is probably equivalent to one cow," he said. Methane emission from cows, through flatulence and belching, varies based on diet, whether the cow is lactating, and other factors. Scientists have estimated that a cow emits between eight and sixteen cubic feet of methane a day.

After the magic trick, Gearhart suggests that we adjourn to the nearby McDonald's for lunch. After eating a small hamburger, Gearhart reaches into his wallet, pulls out a dollar bill, and then counts out 72 cents in change. It's enough for two vanilla ice-cream cones. He offers to buy one for Cooke's grandson, if the young man would stand in line to buy them both. He agreed.

Afterward, we head back to Gearhart's office for a quick tour of the manufacturing side of his business. He introduces us to various workers soldering circuit boards onto small tools and shows off a machine built to calibrate gyroscopes. He makes a magnetic survey tool called the Geo-Shot. It goes down a newly drilled well and reports on the well's direction. States require independent verification of the underground reach of a well. Drilling under a neighbor's property without permission is a century-old problem. It amounts to stealing oil and gas. This practice was dramatized memorably in the 2007 movie *There Will Be Blood*. After years of holding out, a penniless landowner finally offers to lease some land to the deranged oilman portrayed by Daniel Day-Lewis. But the oilman says that he has already drained the land illegally. "I drink your milk shake!" he taunted.

The Geo-Shot prevents illicit milk shake drinking. Texas has had several scandals involving wells drilled at a slant to steal neighboring oil, and since 1949, the state has required the type of directional survey that Gearhart's tool produces.

Making them, Gearhart says, is a good and steady business. At the end of the tour, Cooke takes my elbow and leans in close. "His data is required by regulators," he says. "Someday, if proof was required that a well will not leak, we could supply that proof. It's a long way from being required."

In the months afterward, the lack of interest or demand by the drilling industry in a better and cheaper cement-leak detection tool created headaches for Cooke and Smiley's new Well Integrity Technology Company. State regulators couldn't insist it be used because there were no tools available on the market. Until Cooke and Smiley had a working RDT, they couldn't find a drilling company to test it out.

When the tool was finally located in Bakersfield—the one in the attic turned out to have been loaned years earlier and lost—Smiley hoped that Gearhart could refurbish it. This would let them run it down wells to test out if it was better than a conventional cement bond log. But in early 2013, Gearhart informed him that it would cost $150,000 to design and manufacture a prototype for testing. This was a steep price, said Smiley, because margins are pretty tight in this part of the oil-field service market. And Cooke would have to come up with the money for the prototype himself. If the prototype was lost in the test well, a not-infrequent occurrence, he would have to pay more for another.

Meanwhile, Cooke had thought up a new polymer to be used to frack wells. Compared to standard fracking fluids, it required less water and was biodegradable, so it was more environmentally benign. Promising to lower the cost of fracking a well, this technology had drawn interest from the industry and secured a government research grant.

The RDT tool was going nowhere. As of this writing, Smiley hopes to find a partner—a drilling or oil-field service company—willing to pay for the prototype and run some tests. But interest has been practically nonexistent.

"We're still holding out for some company who is interested in seeing if this works," Smiley said. "We don't have enough time to be knocking on everyone's door."

Would a modern RDT improve the industry's wells? It is possible. As with any invention, it needs a long period of tests and real-world use to determine its usefulness. But Cooke's experience shows that the industry isn't clamoring for better tools, and neither are regulators or landowners. At my urging, my father asked Chesapeake what tests it had used on the well at the Farm. His phone calls were bounced between offices and he never got an answer.

13

PANDORA'S FRACK

On December 1, 2012, Dallas opened a new nature and science museum on the edge of downtown. The building is a light gray cube with an irregularly layered concrete façade that looks like stacked rock strata. From a distance, it appears as if a block of earth has risen out of the ground, shed its soil, and dried out. Entering the building, visitors head toward an escalator that carries them up into the ceiling and then through a cutback in a compressed space surrounded by poured cement walls. The path leads to a fifty-four-foot escalator that climbs past a long diagonal bank of windows. Near the top of the cube, the escalator deposits visitors in a sunlit balcony from which the downtown Dallas skyline is visible.

Rising up, surrounded by cement, a passage through rocks— this passage felt strangely familiar when I visited the museum. A few hours later, it struck me that I had followed the path of a hydrocarbon molecule coming up through the earth, traveling in a man-made opening in the rocks, while hemmed in by concrete. Deposited at the balcony, I had exited the well and entered the modern world with the view of skyscrapers and an eight-lane freeway. Maybe I had too much fracking on the brain. I contacted the architect who designed the Perot Museum of Nature and Science, Pritzker Architecture Prize–winner Thom Mayne. An email from Arne Emerson, an associate of the architect, confirmed that the façade was inspired by underground rock layers, but then suggested that the rest of my impression was the product of an active imagination. "Your interpretation," he wrote, "is one of the best things about experiencing a great piece of architecture—it evokes a response and is also both personal and subjective."

Perhaps, but a lot of people enriched by the recent energy surge gave generously to the museum. A large sign next to the front entrance lists the founding donors. All had oil and gas to thank, to one extent or another, for their wealth. There was a granddaughter of H. L. Hunt, the Texas oil tycoon whose feuding family inspired the television show *Dallas*. Another benefactor was Trevor Rees-Jones, who was drilling conventional wells on land he had leased in and around Fort Worth when the Barnett Shale took him by surprise and made him a billionaire. There was also the former chief executive and chairman of EOG Resources, the second most active driller of wells during the first decade of shale exploration. The final donor, the Perot family, is best known for its computer services company and its patriarch's presidential run. But benefactor Ross Perot Jr. had also signed a lucrative lease to allow Barnett wells on its industrial park and airport north of Fort Worth—and was prospecting in Kurdistan.

This museum was built by energy riches to celebrate engineering triumphs and the natural world around us, prominently including fossil fuels. It was also an attempt to institutionalize and demystify fracking, all wrapped up in a building that will be around to teach future generations about oil and gas, fuels that, as an exhibit inside the museum states, "radically changed the world." Amid a number of smart exhibits on geology and technology, there's a healthy dose of propaganda. Touch a video screen in the Tom Hunt Energy Hall, and a cartoon country-and-western singer launches into an upbeat ditty about the Barnett Shale, accompanied by pictures of jaunty, smiling houses, and dollar signs. The lyrics pair "racket" with "frack it," and rhyme about the importance of paying attention to details while drilling urban wells in Fort Worth. "Take the time to get it right," the singer croons. "There'll be gas tomorrow night."

The song is catchy, but too saccharine. Its description of knowledgeable, patient landmen who "answer questions so it's known they've got competence and skill" isn't in sync with reality. As the land run hit full stride in 2007, most landmen were contractors. The more signed leases they amassed, the more they were paid. Competition among landmen was fierce. Speed and results were valued above all else. The song's sunny mention of smiles "all around"? Some people embraced fracking, while their neighbors viewed the landmen as an invading army.

Even the industry recognized that, at times, it was engaged in a form of ground warfare. At an industry gathering in 2011, an Anadarko Petroleum manager who handled community relations advised the audience to download the US military's counterinsurgency manual "because we are dealing with an insurgency." So which is it? An insurgency? Or smiles all around? In my experience, neither one comes close to capturing the complex reality that unfolded as fracking spread across the country.

When it comes to the domestic drilling boom, common ground is elusive. The forces arrayed in favor and against don't speak the same language. Even the spelling of *frack* is divisive. The November 2008 issue of the *Sylvanian*, a newsletter from the Pennsylvania chapter of the Sierra Club, ran side-by-side letters to the editor. A geologist wrote in support of "fracing," while a worried resident called it "fracking." It was a sign of the brewing linguistic civil war.

The industry has long used *frac*. In 1952 the Stanolind researchers referred to the *hydrafrac treatment*. The industry's preferred spelling remains without the *k*. Its engineers talk of frac jobs, frac fluid, fracing, and fraccing. Critics of the industry almost always say *frack*. Why would opponents use this alternative spelling? My theory is they were, consciously or not, tapping into an existing negative association. In 1978 *frack* appeared as an expletive in the popular science fiction television series *Battlestar Galactica* as a way of allowing writers to sidestep Federal Communications Commission censors. A guide for series writers spelled it *frack*. When the series was resurrected on cable in 2003, the new generation of script writers enthusiastically deployed the word—but spelled it *frak*. They wanted it to be a four-letter word, as in "Frak you," "I don't give a frak," and "You don't want to frak with me." By the time the show ended in 2009, *frak* had taken root in geek culture as a swear word and appeared on popular television shows such as *The Office* and *The Big Bang Theory*. Ron Moore, who developed the second iteration of *Battlestar Galactica*, said he had never heard of the oil-field term. Not that it mattered. By the time opponents started referring to hydraulic fracturing as fracking, they were hitching themselves to an expletive. Before long, signs at protests rallies warned the industry, "Don't Frack with Me." *

---

* I have used the spelling *frack* and *fracking* throughout this book for two reasons. First, they are the preferred spelling of the *Wall Street Journal* and other major newspapers. Second, the spelling *fraced* simply doesn't convey the clipped cadence of the word as it is pronounced by opponents and engineers.

From the Dallas museum, I drove thirty minutes northwest to the suburb of Southlake, an affluent community midway between Dallas and Fort Worth that is home to an inordinate number of current and retired professional athletes. Former Pittsburgh Steelers quarterback Terry Bradshaw lives there. So does current Dallas Cowboys quarterback Tony Romo. Not surprisingly, the local high school team, the Carroll Dragons, has won four state football championships in the last decade.

I wanted to visit Southlake because it tried to take a stand and keep out the rigs. In 2009 John Terrell was elected mayor on a pledge to "preserve Southlake as a great place to live, work, and play." His plan went well for a few months until Exxon Mobil applied to drill a gas well. The ensuing fight shredded the city's self-image as a place of fraternal goodwill and easy access to upscale shopping. "Get the Frack out of Here" signs appeared on well-kept lawns. Two citizens groups emerged: the antidrilling Southlake Taxpayers Against Neighborhood Drilling (STAND) and the pro-drilling Southlake Citizens for Property Rights. The groups sued each other and the city.

Opponents were incensed that gas drilling would be allowed inside the gates. "Everything they've done to try to build this city is going down the tubes," said a protester outside city hall. After a year of acrimony, the anti crowd won a victory when Exxon withdrew its application to drill. Facing the threat of another lawsuit, Mayor Terrell said he had no choice but to allow drilling permits within the city. But the ordinance passed was so stringent as to all but prohibit gas exploration. When I visited, there still had been no drilling. Perhaps the rigs had stayed out due to the new rules or falling gas prices. Maybe Exxon didn't want to stir up a well-financed

and connected hornet's nest a short drive from its corporate headquarters.

I parked my car near one of the proposed drilling locations, a
large gated home on several acres, and went inside a business across
the street that cuts marble for custom kitchen countertops. Marshel
"JR" Melvin, the operations manager, introduced himself. I asked
him why Southlake had managed, so far, to keep fracking at bay.
"They got some serious money here," he said, rubbing his thumb
against two fingers in the international gesture of lucre. "If they don't
want something, it ain't happening." In fact, he continued, there were
neighbors who wanted to see his business depart, since its open shed
with large marble-cutting machines didn't fit the town's image. People here want to live in a bubble that doesn't include any industry,
he said. They wanted their marble countertops but not to see where
the marble was cut and polished. Southlake wanted its large homes
and SUVs but not the machinery of energy exploration.

Southlake, as the name suggests, is on the southern side of
Grapevine Lake. North of the lake, other towns had taken a different
approach, trying to strike a balance that allows drilling but with re-
quirements and conditions. Some of these towns have been working
and reworking their rules for a dozen years, longer than anywhere
else, revising and improving them along the way.

The town of Bartonville, for instance, has a part-time inspector
and requires companies to pay for air and water quality sampling
before and after drilling. Three-story noise-dampening walls around
drill pads, which look like an installation from the artist Christo, are
pretty much compulsory. All work, except drilling, stops between
eight at night and seven in the morning. Fracking can't occur on a
weekend or major holiday. The town requires companies to submit
testing data to city hall to prove wells have been properly cemented
and the well's integrity is solid. Testing of air and underground water

is required before drilling, to establish a baseline, and continues long after the well is complete. Mayor Ron Robertson said he wanted to "hold the fracking company responsible" if something went wrong but also make sure that drilling could take place. "The majority of our citizens are leaseholders," he said, including Robertson and four of the five city council members. There have been no complaints filed about operations there or violations issued by Texas regulators.

By all accounts, the process is working well for the industry and the town's 1,400 residents, which include Rex Tillerson, the chairman and chief executive of Exxon Mobil. There are no wells on his property, but there is one less than a mile from his front door. The well was fracked in 2007. Two years later, Exxon bought the company that owned it. Tillerson doesn't own any mineral rights or benefit from his neighborhood well, according to the company. In this respect, he is similar to Ottis Grimes, the Burkburnett homeowner who sued to stop drilling next to his home. The similarities, of course, end there.

Had Bartonville found the right approach? No set of municipal rules is perfect. One size will not fit all, but residents seemed content. The city keeps changing and updating its rules. It was engaged and willing to learn and, as Mayor Robertson told me, to tighten up loopholes where they appeared. This approach was possible because the town had neither shied away from drilling nor embraced it blindly. The city didn't want to keep out drilling, he said, but felt it had an obligation to keep an eye on energy exploration and insist the highest standards were met. Fracking means the promise and peril of energy production are coming back to the United States, and Bartonville was ready to play its part.

Fifteen years after Nick Steinsberger stood in his boss's doorway and declared the world's first modern frack a success, the future of energy is in some ways brighter than it has been for many years. Fracking has generated an abundance of the energy that society demands and depends on. Wars have been fought over access to energy—simply reducing a fuel subsidy can lead to riots in some parts of the world. We are fossil-fuel addicts. What happens when drug addicts detox? They can be rash, cranky, even psychotic and dangerous. It would be good for the environment if the entire economy abruptly quit fossil fuels, but that's not realistic. I wouldn't want to be around if it ever happened. Perhaps it is best to think of natural gas like methadone. It's a way for an energy-addicted society to get off dirtier fuels and smooth out the detox bumps.

On a smaller scale, there are many people lowering their energy usage. They add insulation and double-paned windows to their homes and purchase more efficient heating and air-conditioning systems. Some run their cars on fuel made from leftover vegetable oil or inedible plant material. Installation of residential solar panels is booming as prices drop, allowing people to generate their own electricity on their roofs. The most committed can even install batteries to store excess solar power and use it at night, cutting the wires that connect them to the grid altogether. This declaration of energy independence—or, at least, independence from fossil fuels—is a personal choice. But it is not a choice that the US economy can make today.

Renewable energy advocates point to Germany as an example of what could be. A quarter of the electricity comes from wind turbines, solar panels, and burning biomass—leftover wood chips and the like. Power generation has also been democratized: the majority of the new renewable electricity comes from homeowners, cooperatives, and municipal governments, not traditional power utilities. Germany pledged to reach the goal of 80 percent renewable power by 2050. It sounds great, and it is—to an extent. It will require new technologies (improved batteries for power storage) and billions of dollars to string up new power lines. Meanwhile, companies that use gas, coal, and nuclear fuel are in trouble. "Conventional power generation, quite frankly, as a business unit, is fighting for its economic survival," said the chief financial officer of RWE, Germany's second-largest utility. Germany began this transition in 1986, after the catastrophic meltdown at the Chernobyl nuclear power plant in Ukraine, and implemented a focused federal energy policy in 2000 to eliminate coal and nuclear power. Energy delivery systems are enormous and complex. Changing them takes time, political will, and patience.

The United States went about its own *Energiewende*—what the Germans call this energy change—driven by a very different set of forces. The United States has stumbled about without a comprehensive energy policy for decades. It relied on the market to supply the energy its people and industries wanted. Early government support for new natural gas extraction technologies led to a much larger free-market investment. Collaboration between shale drillers, millions of landowners, and Wall Street moneymen created a national energy policy: drill a lot of shale wells, unearth a huge new source of natural gas, and let the pieces fall where they may. There was enormous risk-taking by private companies in pursuit of even larger financial rewards.

Perhaps that's a lesson to be learned from both George Mitchell and Aubrey McClendon. You can't have the benefits of a bumper crop of new energy and economic growth without brashly taking chances. Both Mitchell and McClendon set in motion huge technological and social changes without an inkling of understanding where it all would lead. Fracking began as a way to get at the Barnett Shale's natural gas, but has proven just as effective at unlocking the Bakken's crude oil. Rapid drilling and fracking created problems that need to be addressed. Some nations require proof that a new technology will not cause public harm before companies can engage in it. The United States doesn't subscribe to this approach. Exxon's Rex Tillerson, for one, approves. "If you want to live by the precautionary principle, then crawl up in a ball and live in a cave," he said.

The unfettered market is the ugly beauty of the US energy system. You can usually count on this free-market approach to deliver needed energy, but it can also make you want to turn away and not look too closely. It is the responsibility of the industry, landowners, and the energy-consuming public to ask hard questions and make sure there are no corners cut.

Nobody would argue that a nuclear plant should be built as quickly as possible without spending the necessary time to ensure it is safe and robust. Fracking is different. The risks of any single well are tiny compared to a nuclear power plant. But several hundred wells? Several thousand? When McClendon's Chesapeake began to drill and frack thousands of wells, his appetite for risk and ability to push the pace was breathtaking. It is entirely possible for the energy industry to develop oil and gas from shale safely, but not while simultaneously more than doubling the number of wells it drills in a decade.

Fracking arrived at an auspicious time. Just when it seemed like we had used up all the easy-to-tap fossil fuels, forcing the world to get serious about renewable energy technologies, an enormous new supply of oil and gas emerged. It provided an economic boost for the United States—and many large checks to landowners amid the economic meltdown that began in 2008. The US dollar is strengthened by lower imports of crude and energy-intensive industries feasting on cheap gas to export more goods. America's energy revolution is even untangling some of the thorniest foreign policy challenges the nation has faced for the past forty years. Rising US crude oil production could weaken OPEC's influence and make geopolitical demands of major Middle Eastern producers, notably Saudi Arabia, less pressing for Washington's diplomats. OPEC, the supplier of one in three barrels of oil consumed around the world, said in a report posted on its website in April 2013 that the shale-driven oil boom in the United States was "threatening to drastically reduce America's oil-import needs." In 2014 China is expected to pass the United States as the world's largest importer.

In late 2005 the chairman of the Federal Reserve, Alan Greenspan, told Congress that America was running dangerously short of domestic natural gas and warned that rising prices would threaten economic growth. His tutor on the subject was Lee Raymond, then the chairman and chief executive of Exxon Mobil. As it turned out, natural gas prices had a profound impact on the US economy, but not the way that Greenspan predicted. Chesapeake and others found so much natural gas that prices dropped below $4 per million British thermal units—and even dipped below $2 for a brief span. Big industrial consumers of gas scrambled to take advantage of the bounty. Inexpensive gas! In the United States! Steelmaker Nucor, in 2011, began construction of a $750 million iron ore upgrader, creating 150 permanent jobs in Saint James Parish, Louisiana, that paid an average of $75,000 a year. Seven years earlier, Nucor had closed a similar upgrader in the same spot, dismantled it, packed it in crates, and shipped the entire thing to Trinidad, a small Caribbean nation that had discovered natural gas off its shores. Companies that make petrochemicals, fertilizer, and aluminum are building new facilities in the United States, creating a much-needed boost in the midst of the economic downturn created by the real estate crash of 2008 and 2009. After a long decade of hemorrhaging manufacturing jobs overseas, thousands of blue-collar jobs are being created when many believed those jobs were gone for good.

To supply all these new steel mills and chemical plants requires thousands and thousands of new wells. As a result, a major industrialization of both rural and urban parts of the United States is taking place. Wells are being drilled and fracked by rumbling machinery that appear one day in the middle of a field. Following behind are new pipelines and processing plants to strip out ethane, butane, and other valuable liquids from the gas. It would be unwise to throw away this energy and its benefits. It would also be unwise not to seriously grapple with how to extract this energy in a way that lessens the burden for communities above the shale and avoids creating problems for future generations.

I don't fear fracking. I fear carbon. The emergence of fracking—along with improved horizontal drilling and radical improvements in drilling speed and cost—has created an abundance of fuel, but it has also opened a Pandora's box. The availability of all this energy has thrown a lifeline to fossil fuels. There's a great quote attributed to Sheikh Zaki Yamani, the Saudi Arabian oil minister in the 1970s: "The Stone Age did not end for lack of stone, and the Oil Age will end long before the world runs out of oil." Fracking ensures that the age of oil—and its princely hydrocarbon cousin, the natural gas molecule—will not end because we have run out of fossil fuels. But it may end because burning these wonderful fuels puts the planet farther down a path we don't want to head down.

All the economic boosts delivered by increased drilling—and inexpensive energy—won't last forever. As the late petroleum economist Paul Frankel put it succinctly, "There is always either too much or too little." The corollary to Frankel's statement should be: when there is too much, we will use more and more until there is too little again. It's already happening. The Nucor facility is one of a hundred new industrial facilities planned to take advantage of the new abundance of natural gas. Shutting down coal power plants is also increasing demand for gas. Then there are the dozen or so schemes to export natural gas on giant thermos ships. And there are plans afoot to run more cars and trucks on natural gas, and to replace fuel oil for home heating in the northeastern United States. I don't know how—or when—the era of gas abundance will end, but I am sure it will. And given the rush to take advantage of this domestic fuel, I suspect it will be in my lifetime.

Fracking provides a couple decades to make wind and solar generation better, improve battery technology needed to store renewable power for windless, dark nights, and even build new nuclear power plants. Until then, natural gas is the best available option available for reducing carbon emissions, without grinding the wheels of modern economies to a halt.

The shale revolution provides important insights into how energy change takes place, lessons worth studying by people interested in a future powered by renewable fuels. Free-market adherents argue this energy surge was the result of companies making risky investments because of rising fuel prices and private landowners pursuing their economic self-interest. In short, they declare the shale revolution was a victory for market forces. Good government supporters point out that the Energy Department bankrolled early research into shale exploration that was picked up and carried forward by Mitchell Energy. The federal government also provided stable and consistent policies, in particular how the injection of water that is the heart of hydraulic fracturing would be regulated. They contend the government laid the groundwork for the revolution. Finally, there are technocrats who celebrate the power of two disruptive technologies, fracking and horizontal drilling. It wasn't market forces or government policies, they argue, but the power of technology that won the day.

The truth is all three camps are right. To take off, the shale revolution required consistent government policy, disruptive technologies, and a healthy dose of market forces. They all came together and were pulling in the same direction for the first time in decades. The result has been staggering and transformative. Modern societies face many complex challenges, from building an adequate health care system to devising robust economic policy. The lesson of America's energy revolution seems to be that when faced with stubborn problems, the right approach is to broadly align government policy and market forces in order to create fertile ground for disruptive technologies. The growing amount of carbon in the atmosphere creates another seemingly intractable problem. How can a country quickly transition to low-carbon energy without wrecking its economy? The lesson from the development of modern fracking likely applies here as well: create the right market signals, set smart long-term policy goals, and let the technologists develop needed breakthroughs.

It is too soon to declare victory in fracking. The work improving individual wells isn't over. Widespread use of Claude Cooke's leak-detection tool—or something comparable—seems like a smart direction in which to head. It's not the only necessary upgrade. Fracking has expanded so quickly that many eminently sensible technologies to lower the impact of drilling—such as recycling water needed to frack a well or using gas to power the machinery, instead of emission-spewing diesel generators—only began to be adopted years into the drilling boom.

The unofficial policy was to drill first and ask questions later. But questions are being asked and answers are beginning to emerge. How well those answers are heeded will likely determine whether we look back on the shale revolution with relief or regret.

Will the construction of mobile factories for drilling and fracking leave local residents with worse health, whether from breathing
too many air pollutants or being stressed out by the noise, trucks,
and fear of spills? One tool for answering that question is called a
health impact assessment. It's a lengthy process, promoted by the
Pew Charitable Trusts, the Robert Wood Johnson Foundation, and
the US Centers for Disease Control and Prevention, that brings
together local governments, the industry, and residents to tease out
the potential health impacts of an industrial development and devise
ways to mitigate those impacts. The first—and, as of this writing,
the only—assessment of a gas development took place in Battlement Mesa, Colorado, in 2010. This was years after the energy boom
began. The assessment was inconclusive. There weren't enough data,
the public health investigators said, to determine what impact drilling wells nearby would have on the community. Local officials had
to make decisions about permits and were "often unable to wait for
science to catch up," the investigators wrote. Much more research
was needed.

Another basic question about shale development deserves particular attention. Natural gas—specifically its primary ingredient,
methane—is a more potent greenhouse gas than carbon dioxide,
although it breaks down in decades, not centuries. When a shale well
is fracked, the fluid flows back to the surface. It is either recycled or
sent to a disposal well. Some gas comes up with the fluid—and is
often vented into the atmosphere. More gas leaks from pipelines on
its way to residential and industrial consumers. Scientists have been
arguing about whether the typical shale well releases so much methane into the atmosphere that it erases any global warming benefits of
using gas instead of carbon-intensive coal. This question is critical.
Unfortunately, the debate over the subject began in 2010, a full decade into shale development. By that point, the industry had already

drilled one hundred thousand shale wells, and that's a conservative estimate.

What's the answer? A Cornell study, backed by the antifracking Park Foundation, found the amount of natural gas released was so significant, we were better off burning coal. This research was widely criticized, and a subsequent, much larger study by two Massachusetts Institute of Technology researchers came to the opposite conclusion. The debate rages on, although the balance of recent scholarship comes down on the side of the MIT research. It's disturbing how many wells have been drilled before this question of methane leakage was asked. The MIT study points out that a couple states require—and some operators employ—"green" completions. They capture the fluid, take out the natural gas, and sell it. What's striking is that using this green technology costs between $1,000 and $3,000 per day—on wells that can routinely cost $10 million. And by capturing and selling off the gas, a green completion more than pays for itself. This technology will be required by federal rules in 2015.

These upgrades to the basic fracking tool kit could also help reduce air pollution. Natural gas wells, during drilling and fracking, can emit pollutants and chemicals that form ozone. Indeed, a rural corner of Wyoming (population 11,500) with substantial drilling has recorded concentrations of ozone-forming pollutants similar to metropolitan Chicago (population 9.2 million) and Phoenix (population 3.8 million). Breathing in high levels of ozone can impair lung function and trigger asthma. Even the Texas Commission on Environmental Quality, not known for its aggressive pursuit of industry, found instances when faulty valves released enough benzene in the Barnett Shale to exceed air-quality standards.

Why wouldn't the industry adopt technologies that both protect the environment and boost its bottom line? The same reason that there are enormous flares of natural gas in North Dakota. The industry has wanted to move extremely quickly to drill the wells—to beat the ticking clock of lease obligations, to meet or exceed Wall Street's earnings forecasts, and to begin generating a return on the money it invested in wells. In North Dakota, companies don't want to wait for gas pipelines to be installed, not when it can begin to pump out valuable crude. And most don't want to wait around if green completion equipment isn't available.

It's time to slow down. My father rewired the Farm—even if the inspector's complaint was that the electrical boxes were a quarter inch too small. "If anything happened to those kids because of the wiring, I would never forgive myself," he said. It might not have made any difference, but maybe it would have. When it comes to fracking, getting it right is important. And if we blow it, we're never going to forgive ourselves.

So, as the animated country singer in the Dallas museum intoned, "Take the time to get it right. There'll be gas tomorrow night."

Thanks and Acknowledgments

When I began reporting on the petroleum industry more than a decade ago, my knowledge about fossil fuels was limited. Gasoline came in three flavors: regular, premium, and superpremium. It had something to do with crude oil. I thought oil and gas was a low-tech business. I once naïvely asked a sales representative from a data storage company what he was doing at the world's largest gathering of oil vendors. Computing power was for Silicon Valley, not Houston.

I owe an enormous debt to many people who over the years have shared their knowledge of the industry with me, so that I could understand the modern energy industry in all its glory and foibles. Countless petroleum engineers took time explaining the intricacies of how to find oil and gas, to drill a well, and to hydraulically fracture it. Their patience with my generalist's understanding of science was a gift. Geologists provided informal seminars on the nature of hydrocarbon reservoirs and shale. Critics of fossil fuels also provided invaluable assistance to help me see the industry through their eyes. I hope that

I repaid their time by accurately describing their work and views. All mistakes are mine.

Many oil companies regard reporters as pests to be managed. Marathon Oil was very generous to allow me to witness the fracking of the Irene Kovaloff. Thanks to the folks in the Dickinson field office for their willingness to throw their schedules to the wind.

My professional home during this time was the *Wall Street Journal*'s Texas bureau. I have been fortunate that the bureau has been peopled with a string of great editors and colleagues. I couldn't begin to repay the support and guidance offered by Leslie Eaton. Before her, Jennifer Forsyth provided early encouragement. Karen Blumenthal, who suggested I try the energy beat back in 2002, deserves recognition and my sincere appreciation for teaching me, among other lessons, how to read a cash flow statement. Ben Casselman and Daniel Gilbert covered Chesapeake after me. Their reporting on and insights into the company were top-notch. Angel Gonzalez, who for years was our man in Houston, has been a valued coworker and friend. While never a Texan, Bhushan Bahree generously shared his sagacity derived from decades covering the world of oil. There are few editors as smart as Mike Williams, who served for a time as the paper's energy czar.

Many thanks to the University of Texas for keeping its stacks open to the public, even those of us who married Aggies. I benefited tremendously from access to the collections of the Perry-Castañeda Library, the McKinney Engineering Library, the Walter Geology Library, and the Dolph Briscoe Center for American History. In addition, thanks to folks at the public libraries in Oklahoma City and Tulsa for their help, especially Sheri Perkins at the Tulsa City-County Library, who went above and beyond to find an old newspaper article I needed. Computers make so much information available, but sometimes nothing beats a well-maintained vertical file. Also deserving specific thanks are Mat Darby at the Briscoe Center and Rebecca Radford at the Kansas Geological Society & Library.

Many have earned my gratitude for their willingness to engage in meandering discussions of energy. At the top of this list is Professor Michael Webber of the University of Texas. Why he orders Mexican plates at a Brazilian restaurant is a mystery, but his deep understanding of energy isn't.

To Elizabeth Gold, thanks for your sanity, wise counsel, funny stories, insights, observations, and vigilance keeping tabs on the Park Slope antifrack pamphleteers. For your main squeeze, Danny "Inspector" Felsenfeld, here's another literary mention to add to your collection.

This book wouldn't have been possible if David McCormick hadn't seized upon my idea with gusto. Among his many contributions was steering it into the hands of Ben Loehnen, a most able editor.

Over the past couple years, I have told people this book began as a birthday present to myself. The present was permission to spend long hours alone, hashing out ideas, and writing drafts. I now realize that I didn't give that present to myself. My wife and children were the gift givers. Their love, unstinting support, and encouragement were constant and appreciated in ways I can never fully express. The book is dedicated to them, but that doesn't come close to repaying them. Thanks.

Sources

Below are listed the primary sources, by chapter, I relied on to research and write this book.

Chapter 1: Just Add Water
Chesapeake reported, in its annual 10-K report filed in March 2013, that it held leases, either partial or whole, on 25.9 million acres. This is about 40,500 square miles. Kentucky covers 40,400 square miles. In 2012 the company reported expenditures of $21.6 billion.

The history of the Farm comes from several talks with my parents, Barbara and Steve Gold, as well as records in the Sullivan County courthouse in Laporte, Pennsylvania. The earliest prominent reference to a "Shale Gas Revolution" is a 230-page research report written in part by Chris Theal and issued by Tristone Capital Global on October 6, 2008. Details on George Mitchell and Mitchell Energy are included in the sources for chapter 5. More information about Oklahoma City in the 1980s oil boom and bust can be found in Mark Singer's wonderful book *Funny Money*. I assembled information on the Matt 2H well from many sources, including the Susquehanna River Basin Commission, the Pennsylvania Department of Environmental Protection, and FracFocus Chemical Disclosure Registry (www.fracfocus.org). I tried, unsuccessfully, to track down the electrical inspector who refused to issue a connection permit through the Sullivan County Rural Electric Cooperative.

Figuring out how many wells are fracked annually isn't as easy as it would seem. I relied on Richard Spears of Spears & Associates, publisher of the annual and authoritative *Oilfield Market Report*.

The best source of data on national energy consumption on a per-capita basis is the World Bank's calculation of fossil fuel energy use. I relied on its "World Development Indicators: Energy Production and Use" table, which can be found here: http://wdi.worldbank.org/table/3.6#. (Last accessed August 2013.) US per-capita consumption of fossil fuel is about 10 percent greater than Canada. Other nations with higher per-capita fossil fuel consumption are major oil producers, such as Qatar and Kuwait, and use the large amounts of fuel to power export-oriented industries. The International Energy Agency's *World Energy Outlook 2012* is the source of the claim that by 2020 the United States may become the world's largest oil producer and for the relative carbon emissions of coal, natural gas, and crude oil.

The section on global shale deposits and how shale was formed was aided greatly by numerous conversations and email exchanges with: Julio Friedmann, the chief energy technologist at Lawrence Livermore National Laboratory; Scott Tinker, director of the Bureau of Economic Geology at the University of Texas; Juergen Schieber, "mudman" and "shale detective," and, more conventionally, a geology professor at Indiana University; James Coleman, a research scientist at the United States Geological Survey; and Ray Levey, director of the Energy & Geoscience Institute at the University of Utah. I consulted several geology papers and found work by Gregory Ulmishek and Richard M. Pollastro particularly helpful. In addition, Clint Oswald and his colleagues at Wall Street research firm Sanford C. Bernstein have published many great notes on global shale deposits, including "Oil Shale—Forget About the Bakken, Wait for the Bazhenov" in 2012.

Birol, Fatih. *World Energy Outlook 2012*. Paris: IEA Publications, 2012.

Deutch, John. "The Good News About Gas: The Natural Gas Revolution and Its Consequences." *Foreign Affairs* 90, no. 1 (January–February 2011): 82.

Gibson, Arrell Morgan. *Oklahoma: A History of Five Centuries*. 2nd ed. Norman: University of Oklahoma Press, 1981.

Medlock III, Kenneth B., Amy Myers Jaffe, and Peter R. Hartley. *Shale Gas and U.S. National Security*. Houston: James A. Baker III Institute for Public Policy, Rice University, 2011.

Sinclair, Upton. *Oil!: A Novel*. New York: Penguin Books, 2007.

Singer, Mark. *Funny Money*. New York: Knopf, 1985.

Chapter 2: Ottis Grimes

I relied on several sources for the history of Burkburnett, Texas. These included three books, cited below, as well as an oral history of Walter Cline in the University of Texas's Dolph Briscoe Center for American History, Pioneers in Texas Oil archive, and the Handbook of Texas Online (http://www.tshaonline.org/handbook/online/articles/heb14) entry on Burkburnett, last accessed October 2013.

Details about Ottis Grimes came from court decisions in the case. The 78th District Court in Wichita Falls, Texas, was kind enough to pull the original case file and send me a copy. This is the source of the "commonly sustained" quote. Another particularly helpful source was the Court of Civil Appeals (Fort Worth, Texas) decision in *Grimes v. Goodman Drilling Co., et al.* (no. 9213), issued June 14, 1919, and cited in the *Southwestern Reporter,* vol. 216 (St. Paul: West, 1920), 202–4. The Barney Fudge quote is from an interview with the author. For a modern example of homeowners without mineral rights fighting oil companies, see the *Ruggiero v. Aruba Petroleum* lawsuit, 271st Judicial Court (Wise County, Texas).

The Rex Tillerson quote is from a transcript of his meeting with the *Wall Street Journal* editorial board that was provided to me by Exxon and used by permission. A partial quote first appeared in an article I cowrote that appeared in the December 3, 2012, edition of the *Wall Street Journal:* "Global Gas Push Stalls—Firms Hit Hurdles Trying to Replicate U.S. Success Abroad."

The quote from Shell's Peter Voser was in an October 2012 interview with the author. John Tintera's reflections on his tenure at the Texas Railroad Commission are from a panel discussion convened by the Alaska Oil & Gas Congress and quoted in a September 22, 2013, post by Starr Spencer in Platts's *The Barrel* blog. The Emily Krafjack quotes are from an interview with the author, as is the "industry has done a great job of figuring out how to crack the code" quote from Mark Boling. The data on job growth is from Stephen P. A. Brown and Mine K. Yücel's *The Shale Gas and Tight Oil Boom: U.S. States' Economic Gains and Vulnerabilities,* a Council on Foreign Relations report from October 2013. The sausage quote was cited in a *Charleston Gazette* article from March 22, 2012, "Gas Drilling Needs to Improve, Chesapeake Official Says," by Ken Ward Jr.

I referred to studies in Pennsylvania, conducted near my parents' property, that looked at the potential for fracking to create pathways for briny water and chemicals to migrate upward into shallow aquifers. These are the Warner and Osborn papers cited below. The "unlikely" quote is from the

2011 paper and was repeated in the 2012 Warner paper. The use and problems of sewage plants to clean up flowback water is from the 2013 Warner paper, cited below, and correspondence with coauthor Robert Jackson.

The Ernest Moniz quote is from his July 2011 testimony before the Senate Energy and Natural Resources Committee and also from a speech he gave, which I attended, in December 2012 at the University of Texas. The Energy Information Administration provided information about Texas and North Dakota crude oil production.

Data on Nigerian oil imports to the United States comes from the US Energy Information Administration's "U.S. Imports by Country of Origin." The IEA also provides statistics on the nation's output since 1980. I derived the value of its oil output from these data.

There have been several studies examining how renewable and gas generation can coexist. The study I relied on is titled "Partnering Natural Gas and Renewables in ERCOT," carried out by the Brattle Group for the Texas Clean Energy Coalition and was published in June 2013. While a bit technical, it is the most thorough examination of how the power grid currently mixes various generation sources.

A basic overview of the fracking process can be found in two Society of Petroleum Engineers papers by George E. King, who was also generous with his time. The same can be said for Stanford University professor Mark Zoback, who has also given several lectures on fracking that are available online. The quote in this chapter is from a symposium on October 5, 2011.

The rapid growth of shale gas production comes from *The Shale Revolution,* an extensive research report for investors from Credit Suisse and issued on December 13, 2012. It uses cubic meters, which I converted to cubic feet using a 1:35 ratio. I calculated wind's contributions to US power generation using the Energy Information Administration's *Electric Power Monthly* data, comparing trailing twelve-month data from July 2013 to annual data from 2003.

The nonprofit online news operation ProPublica has tracked spending by state regulators. Its February 2013 report on the subject, "State Oil and Gas Regulators Still Spread Thin," has useful data on the topic. It can be found at www.propublica.org/article/update-state-oil-and-gas-regulators -still-spread-thin. (Last accessed August 2013.)

Graf, Edwin A. "Reasonable Use—Determination Whether a Land Use by a Mineral Lessee is Reasonably Necessary Requires Consideration of Alternate Methods of Development Available to a Lessee and a Surface Owner." *Texas Law Review* 50 (April 1972): 806–813.

House, Boyce. *Oil boom: The Story of Spindletop, Burkburnett, Mexia, Smack-over, Desdemona, and Ranger.* Caldwell, ID: Caxton Printers, 1941.

King, George E. "Thirty Years of Gas Shale Fracturing: What Have We Learned?" Paper presented at SPE Annual Technical Conference and Exhibition, Florence, Italy, September 19–22, 2010.

———. "Hydraulic Fracturing 101: What Every Representative, Environmentalist, Regulator, Reporter, Investor, University Researcher, Neighbor and Engineer Should Know About Estimating Frac Risk and Improving Frac Performance in Unconventional Gas and Oil Wells." Paper presented at SPE Hydraulic Fracturing Technology Conference, the Woodlands, Texas, February 6–8, 2012.

Landrum, Jeff. *Reflections of a Boomtown: A Photographic Essay of the Burkburnett Oil Boom.* Burkburnett, TX: Self-published, 1982.

Osborn, Stephen G., Avner Vengosh, Nathaniel R. Warner, and Robert B. Jackson. "Methane Contamination of Drinking Water Accompanying Gas-Well Drilling and Hydraulic Fracturing." *Proceedings of the National Academy of Sciences of the United States of America* 108, no. 20 (May 17, 2011): 8172–76.

Vallon, Dick. *Burkburnett: It's Only the Beginning.* Virginia Beach, VA: Donning Co. Publishers, 2007.

Warner, Nathaniel R., Robert B. Jackson, Thomas H. Darrah et al. "Geo—chemical Evidence for Possible Natural Migration of Marcellus Formation Brine to Shallow Aquifers in Pennsylvania." *Proceedings of the National Academy of Sciences* of the United States of America 109, no. 30 (July 24, 2012): 11961–66.

———. "Reply to Engelder: Potential for Fluid Migration from the Marcellus Formation Remains Possible." *Proceedings of the National Academy of Sciences* of the United States of America 109, no. 52 (December 26, 2012): E3626.

Warner, Nathaniel R., Cidney A. Christie, Robert B. Jackson, and Avner Vengosh. "Impacts of Shale Gas Wastewater Disposal on Water Quality in Western Pennsylvania. *Environmental Science & Technology.* Published online September 2013.

Chapter 3: Everyone Comes for the Money

Marathon Oil, and in particular, Pat Tschacher and John Porretto, made it possible for me to visit the Irene Kovaloff well. Additionally, Lance Langford at Statoil and Bud Brigham helped me understand what the Bakken was all about. Travis Kelly, of Target Logistics, gave me a tour of one of his company's man camps and is also the source of the claim that it will soon house one of every one hundred North Dakotans.

The Bill Klesse quote comes from an interview with the author on August 20, 2012. Details of North Dakota oil production come from both the EIA and the *Director's Cut*, a monthly publication of data from Lynn Helms, director of the Oil and Gas Division at the North Dakota Department of Mineral Resources. In an August 2010 presentation, the department estimated that there would be twenty thousand Bakken wells. By late 2012, it was estimating forty-five thousand, according to a conversation with Alison Ritter, a department spokeswoman. Further information on global and US oil production, and import and export levels, comes from both the IEA and EIA, in particular the EIA's data files with the following source keys: MCRNTUS2, MCRFPUS2 and MCRIMXX2.

I have talked to the estimable Phil Verleger, an energy economist with few peers, for several years, and he was helpful in thinking about the Bakken and its potential impact. The quote in the chapter comes from a conversation and first appeared in a November 17, 2011, *Wall Street Journal* article. He was also a source of my thinking about a long-term link between rising oil prices and economic contraction. In addition, University of California, San Diego, professor James D. Hamilton has written about this.

The quote from Dave Lesar about the "strategic guar reserve" came from a Halliburton conference call with investors on July 23, 2012. The *Denver Business Journal* reported on John Hickenlooper and Lesar drinking frack fluid on March 8, 2012, in a posting on its website titled "Are you drinking the Kool-Aid about fracking fluid?" His testimony to the US Senate Committee on Energy and Natural Resources was on February 12, 2013.

Quotes from Energy Secretary Chu are from a speech at the CERAWeek conference in March 2010, seen by the author. A video of the speech was provided to me by IHS, the parent company of industry consultant CERA. Rex Tillerson talked to the author about his wife's present in a meeting several years ago at the *Wall Street Journal*. His quote about the hearse he expects to take him to his funeral come from a *Financial Times* article on June 27, 2007, "Exxon Chief Sceptical of Plans to Scale Up Biofuels Production," by Ed Crooks and Fiona Harvey.

Details about the Olson 10-15H well come from a report by Neset Consulting Service that I found in the North Dakota Industrial Commission, Oil and Gas Division, records in Bismarck, North Dakota. Details about Brigham Exploration's dire financials come from Bud Brigham and are reported in the 10-K filed by the company on March 13, 2009.

I determined average wages by using state US Bureau of Labor Statistics data and was guided by Michael Ziesch, manager of the Labor Market Information Center of Job Service North Dakota.

Details on the preliminary production of the Irene Kovaloff provided by Marathon Oil.

Hamilton, James D. "Oil and the Macroeconomy Since World War II." *Journal of Political Economy* 91, no. 2 (April 1983): 228–48.

———. "Causes and Consequences of the Oil Shock of 2007–08." *Brookings Papers on Economic Activity* (Spring 2009): 215–59.

Nordeng, Stephan. "A Brief History of Oil Production from the Bakken Formation in the Williston Basin." *Geo News* (January 2010): 5.

Rankin, R., M. Thibodeau, M. C. Vincent, and T. T. Palisch. "Improved Production and Profitability Achieved with Superior Completions in Horizontal Wells: A Bakken/Three Forks Case History." Paper presented at SPE Annual Technical Conference and Exhibition, September 19–22, 2010, Florence, Italy.

Sorensen, James A. *Evaluation of Key Factors Affecting Successful Oil Production in the Bakken Formation, North Dakota.* Assessment presented to National Energy Technology Laboratory, Morgantown, WV, 2008.

Chapter 4: Dominion over the Rocks

I relied on many sources for this chapter, the most important of which are listed here. I found information about Edward Roberts in an 1864 edition of *Dental Times: A Quarterly Journal of Dental Science*; see the patent notice on page 180, as well as "Improved Vulcanizing-Machine," Letters Patent 37,523 (January 27, 1863), and "Improvement in Apparatus for Vulcanizing Rubber, &c," Letters Patent 23,948 (dated May 10, 1859.) The torpedo patent is 59,936 (November 20, 1866). His hiring of Pinkerton detectives is from the American Oil & Gas Historical Society's online article "Shooters—A 'Fracking' History" and is available at http://aoghs.org/technology/shooters-well-fracking-history. (Last accessed August 2013.) The description of the "groan like a great monster" is from the February 4, 1865, edition of the *Boston Commercial Bulletin*. Paul Adomites shared with me a prepublication version of his article "The First Frackers—Shooting Oil Wells with Nitroglycerin Torpedoes," which later appears in *Oil-Industry History* 12, no. 1 (December 2011).

Details about acidization and other twentieth-century techniques come from a number of sources, including a November 1938 *Popular Mechanics* article, "New Oil Wells from Old Ones," and, in the magazine's December 1942 issue, "Bringing Old Oil Wells Back to Life." The steel shortage and post–World War II oil demand and shortages are described in the February 9, 1948, issue of *Life* in "The U.S. Runs Short of Oil." The cost of Stanolind's research center was reported in the February 1953 issue of *Resourceful Oklahoma*, a publication of the Oklahoma Planning and Resources Board. The description of Stanolind's facility is in "Labs Bring City Annual Payroll of $11 Million," in the *Tulsa World*, March 8, 1961.

The Stanolind researchers published a number of scientific papers and patents. Bob Fast's and George Howard's 1970 book on hydraulic fracturing was also extremely helpful. Rebecca Radford at the Kansas Geological Society & Library in Wichita helped me find the scouting tickets for the Klepper #1 well, and historian Larry Skelton helped me decipher them. Details of the Klepper well can be found on pages 8 and 9 in Howard's and Fast's book. Bob Fast's son Rob helped me with the family history. The Stanolind researchers filed two patents nearly simultaneously that established the basics of midcentury fracking. Floyd Farris is listed as the inventor on Patent 29,922 (later reissued as Patent 23,733). The same day, J. B. Clark filed a nearly identical patent. Both patents were filed on behalf of Stanolind. As best I can determine, the two patents vary in one small detail having to do with the orientation of fractures, and Stanolind likely filed both to protect itself. Key papers include: Floyd Farris's "Method for Determining Minimum Waiting-on-Cement Time," in *Petroleum Technology* (January 1946); George Howard's and Bob Fast's "Squeeze Cementing Operations" in *Petroleum Transactions, AIME* (vol. 189, 1950); J. B. Clark's "A Hydraulic Process for Increasing the Productivity of Wells" in *Journal of Petroleum Technology* (vol. 1, 1949); and Bob Fast's "A Study of the Permanence of Production Increases Due to Hydraulic Fracture Treatment" in *Petroleum Transactions, AIME* (vol. 195, 1952). Another important early paper is M. King Hubbert's and David G. Willis's "Mechanics of Hydraulic Fracturing" in *Petroleum Transactions, AIME* (vol. 210, 1957). Hubbert famously spelled out his peak theory in "Nuclear Energy and the Fossil Fuel" in *Drilling and Production Practice* (1956). The Hubbert quote is from a 1989 oral history found here: http://www.oil crisis.com/hubbert/aip/aip_vi.htm. (Last accessed August 2013.)

Details of the fatal explosion in Tulsa were difficult to find. Sheri Perkins at the Tulsa City-County Library did yeoman's work digging up the November 12, 1970, edition of the *Tulsa World*, where the article "8 Die in Blast Near Tulsa" by George Wesley and Pat Crow lay out the details.

President Nixon's "Special Message to the Congress on Energy

Resources" was delivered on June 4, 1971, and can be found online through the University of California, Santa Barbara, at this web address: www.presidency.ucsb.edu/ws/?pid=3038#axzz1l0Kk8Ot8. (Last accessed August 2013.) His quote "My doctor tells me" is from his "Address to the Nation About Policies to Deal with the Energy Shortages" and was delivered on November 7, 1973. It can be found at http://www.presidency.ucsb.edu/ws/?pid=4034. (Last accessed August 2013.)

The "something we can use any day of the week in any gas field" quote is from "Project Gasbuggy" in the September 1967 issue of *Popular Mechanics*. Details about Project Rio Blanco and the plaque are from the Department of Energy's "Rio Blanco, Colorado, Site Fact Sheet" issued in October 2011 and available at www.lm.doe.gov/WorkArea/linkit.aspx?LinkIdentifier=id&ItemID=1006. (Last accessed August 2013.) The "Gas from Project Rio Blanco Unit Could Equal 10 Year Supply" quote is from Frank Kreith and Catherine B. Wrenn's book, cited below.

"When everything else fails, frac it" was quoted in Ahmed S. Abou-Sayed's preface to the third edition of *Reservoir Stimulation*, cited below.

Details of the work done by Al Yost and others at the Morgantown Energy Technology Center is available on two CDs. In 2007 the center published *DOE's Unconventional Gas Research Programs 1976–1995: An Archive of Important Results*, which is a good place to start for those interested in learning more. A couple of the more important papers are listed below.

Economides, Michael J., and Kenneth G. Nolte. *Reservoir Stimulation*. 3th ed. Chichester, UK: John Wiley & Sons, 2000.

Frehner, Brian. *Finding Oil: The Nature of Petroleum Geology, 1859–1920*. Lincoln: University of Nebraska Press, 2011.

Giddens, Paul H. *The Birth of the Oil Industry*. New York: Macmillan Co., 1938.

———. *Early Days of Oil: A Pictorial History of the Beginnings of the Industry in Pennsylvania*. Princeton, NJ: Princeton University Press, 1948.

History of Petroleum Engineering. New York: American Petroleum Institute, 1961.

Howard, George C., and C. Robert Fast. *Hydraulic Fracturing*. New York: Society of Petroleum Engineers of AIME, 1970.

Kreith, Frank, and Catherine B. Wrenn. *The Nuclear Impact: A Case Study of the Plowshare Program to Produce Gas by Underground Nuclear Stimulation in the Rocky Mountains*. Boulder, CO: Westview Press, 1976.

Lederer, Adam. "Using Public Policy Models to Evaluate Nuclear Stimulation

Projects: Wagon Wheel in Wyoming." Master's diss., University of Wyoming, 1998.

McLaurin, John J. *Sketches in Crude-Oil: Some Accidents and Incidents of the Petroleum Development in All Parts of the Globe.* Harrisburg, PA: self-published, 1896.

Moniz, Ernest, Henry D. Jacoby, Anthony J. M. Meggs, et al. *The Future of Natural Gas: An Interdisciplinary MIT Study.* Cambridge, MA: MIT Energy Initiative, Massachusetts Institute of Technology, 2011.

Morra, Frank, J. P. Brashear, and Mark R. Haas. "Comparative Economics of Gas Production from Conventional, Tight, and Deep Reservoirs." Paper presented at SPE Unconventional Gas Recovery Symposium, Pittsburgh, May 16–18, 1982.

Tarbell, Ida M. *The History of the Standard Oil Company (Briefer Version).* Edited by David M. Chalmers. New York: Harper & Row, 1966.

Yergin, Daniel. *The Prize: The Epic Quest for Oil, Money, and Power.* New York: Free Press, 1991.

———. *The Quest: Energy, Security and the Remaking of the Modern World.* New York: Penguin Press, 2011.

Yost II, A. B., W. K. Overbey Jr., and R. S. Carden. "Drilling a 2,000-ft Horizontal Well in the Devonian Shale." Paper presented at SPE Annual Technical Conference and Exhibition, Dallas, September 27–30, 1987.

Yost II, A. B., and W. K. Overbey Jr. "Production and Stimulation Analysis of Multiple Hydraulic Fracturing of a 2,000-ft. Horizontal Well." Paper presented at SPE Gas Technology Symposium, Dallas, June 7–9, 1989.

Chapter 5: Wise County

I am greatly indebted to Todd Mitchell for helping me interview his father, George Mitchell. He coordinated the meeting and accompanied me to help keep his father focused, but he also allowed me to persistently pepper him with questions. Todd Mitchell provided me information about the family's financial straits and health problems in the mid to late 1990s. John C. Todd, son of Cambe Blueprint owner John Todd, was also helpful in recalling details of the office and his father's relationship with George Mitchell. I interviewed several others who worked for Mitchell Energy, including Steve McKetta and Dan Steward, both of whose recollections were helpful. University of Texas professor Jurgen Schmandt shared several interviews he conducted with George Mitchell. This is the source of the "Smart lawyers would convince a jury of anything" quote. I also reviewed many years of public statements and SEC filings from Mitchell Energy. The Texas Independent

Producers and Royalty Owners Association (TIPRO) oral history archive, and specifically the Mitchell interview, at the Dolph Briscoe Center for American History at the University of Texas, was a rich source of detail.

Tonia Wood of the Texas State Library and Archives Commission helped me find numerous articles and correspondence from the 1990s. The material is located in the central correspondence files of the Records of Texas Governor George W. Bush, Archives and Information Services Division, Texas State Library and Archives Commission. This is the source of Carrie Baran's "I am also pro business" letter. Another important document is a July 20, 1998, memo from Mike Regan, the Texas Railroad Commission's executive director, to Governor Bush's executive assistant, Joe Allbaugh. Dennis Meadows shared with me what he remembered about his work with George Mitchell. I also relied on the *Time* article "Environment: The Worst Is Yet to Be?" from the January 24, 1972, edition.

Details on the birth of petroleum engineering are from the 1961 *History of Petroleum Engineering* cited above. The surprising detail that two of the first three people to receive this degree were Chinese came from this book. They are Chun Young Chan and Barrin Ye Long. The third graduate was Frederick Arthur Johnson. The quote "It appears that this technique is probably not economically feasible" is from a 1977 report issued by the Energy Research and Development Administration's Bartlesville Energy Research Center titled *An Application of MHF Technology to a Tight Gas Sand in the Fort Worth Basin.*

Another source of information was the Texas Railroad Commission's records. Details of the 1982 C. W. Slay #1 well are in Oil and Gas Dockets nos. 5, 7B and 9–80, 413. The 1977 and 1978 investigation into Wise County water and Mitchell drilling practices can be found in Railroad Commission Oil and Gas Docket no. 9-68644, located in the records room of the Commission in Austin. This is where I found Darwin K. White's letter. He filled in the rest of the story in an interview. The "probable source" finding comes from the December 16, 1977, commission report titled "Results of Investigation into Gas Contamination of Lower Trinity Aquifer." I found the state's subsequent investigative material in the commission's files under Oil and Gas Docket no. 09-0218133. This lengthy file contains the history of this regulatory action. Details of Mitchell Energy's letter proposing lowering the state fine and response are dated November 23, 1998, and the settlement agreement is dated December 21, 1998. Information about the D. J. Hughes well can be found at the Texas Railroad Commission production records for API well no. 49780215.

The lawsuit against Mitchell Energy is well documented, if largely forgotten. There were the court files and the Texas Court of Appeals appellate decision, case no. 2-96-227-CV (November 13, 1997). Bill Keffer and Randy Miller both shared their recollections and files with me. Miller's quote that Mitchell saw Wise County as "a cash register" came from a February 2012 interview with the author. I also relied on a number of contemporary newspaper and magazine articles, including "Flaming Justice" by Janin Friend in the January 1997 issue of *Texas Business*, which is the source of the "burned-up, parched, miserable place" quote. The *Dallas Morning News*, *Houston Chronicle*, and Associated Press covered this case closely.

Also thanks to Robert Mace and Charles Kreitler, a former Mitchell Energy hydrogeologist, for talking Texas water and methane pollution with me.

Childs, William R. *The Texas Railroad Commission: Understanding Regulation in America to the Mid-Twentieth Century.* College Station, TX: Texas A&M University Press, 2005.

Kutchin, Joseph W. *How Mitchell Energy & Development Corp. Got Its Start and How It Grew: An Oral History and Narrative Overview.* Universal Publishers, 2001.

Meadows, Dennis L., ed. *Alternatives to Growth-I: A Search for Sustainable Futures: Papers Adapted from Entries to the 1975 George and Cynthia Mitchell Prize and from Presentations Before the 1975 Alternatives to Growth Conference, Held at the Woodlands, Texas.* Cambridge, MA: Ballinger Pub. Co., 1977.

Meadows, Donella H., Dennis Meadows, and Jorgen Randers. *Limits to Growth: The 30-Year Update.* 3rd ed. White River Junction, VT: Chelsea Green Publishing Company, 2004.

Schmandt, Jurgen. *George P. Mitchell and the Idea of Sustainability.* College Station, TX: Texas A&M University Press, 2010.

Steward, Dan B. *The Barnett Shale Play: Phoenix of the Fort Worth Basin—A History.* Fort Worth, TX: Fort Worth Geological Society, 2007.

Chapter 6: Ice Doesn't Freeze Anymore

This chapter is based mostly on recollections of people involved and well records. Nick Steinsberger graciously spent hours emailing back and forth with me as we, through a process of elimination and detective work, figured out which was the world's first successfully fracked well. Then he spent a day with me to visit the well and share his recollections.

The S. H. Griffin is API well no. 42-121-30670. There are numerous helpful paper and electronic records about the well in Texas Railroad Commission files. Ray Walker and Mark Whitley spoke to me at length, as did others active in Fort Worth in the late 1990s. Ray Walker and Nick Steinsberger (and four others) collaborated on "Proppants, We Still Don't Need No Proppants—A Perspective of Several Operators," a 1998 paper presented at SPE Annual Technical Conference and Exhibition in New Orleans. In addition to being humorously titled, the paper was very helpful to understanding the thinking of completion engineers at the time. This is the source of the "Why it works is still generally unknown" quote.

Jean-Philippe Nicot's *Oil & Gas Water Use in Texas: Update to the 2011 Mining Water Use Report* was useful to understand fracking water consumption issues. Dr. Nicot works for the Bureau of Economic Geology, and the report was prepared for the Texas Oil & Gas Association.

Chapter 7: Larry Was the Brake

For the history of Devon, I relied on Bob Burke's book on the life of John Nichols as well as Larry Nichols's recollections from a 1996 speech preserved by the Newcomen Society for the Study of the History of Engineering and Technology. The Burke book provides an interesting glimpse of how raising money for domestic energy exploration was professionalized and is the source of the quote about John being the accelerator and Larry the brake.

I also pored through numerous SEC filings from both Mitchell Energy and Devon Energy to better understand the companies. The form S-4 (filed by Devon on August 30, 2001) and later amendments to the original provide information about the merger, which was fleshed out by my interviews with participants. Devon's description of the Barnett as "unique" can be found in its 10-K filed in March 2003.

At Devon, Jeff Hall was particularly helpful, as were Vince White and Chip Minty, who helped me line up interviews with Larry Nichols, Brad Foster, Tony Vaughn, and Jeff Agosta. I tracked down others, including Chance Jackson and Jay Ewing, on my own. Jay keeps several years of his work calendars and helped me pinpoint the date that I first toured the Barnett. Sarah Fullenwider and Fort Worth's approach to regulation come through interviews with her and other city officials.

While working for the *Wall Street Journal*, I obtained (and thankfully filed away) the June 2002 Devon presentation "Mid-Year Operations Update & Barnett Shale School." My interviews with Dick Lowe, Trevor Rees-Jones, and others were conducted for a story I reported several years ago. Information about Hallwood Group and Bill Marble's quote about learning to respect, not fear, the Ellenberger formation is from a presentation he made to the Ellison Miles Geotechnology Institute, and details of his speech to the Fort Worth Geological Society are in the group's January 2004 newsletter. Keith Hutton's reaction is from a 2005 interview with the author.

I reviewed information about Devon's Veale Ranch #1H and Graham Shoop #6 wells, API no. 42-439-3041200 and no. 42-121-3168500, through the Texas Railroad Commission website. I also consulted records for the Lakeview #1H, the well that opened Aubrey McClendon's eyes to the potential of the Barnett Shale. His spoke about this well on a December 1, 2004, conference call and again at an energy conference on June 1, 2009.

Professor Mark Zoback, at Stanford, relayed the story about the informal survey of engineers and provided me with the results of the survey.

The story of former University of Texas bandmates Jack Randall and Rex Tillerson making a deal was reported in Brian O'Keefe's April 2012 article in *Fortune* magazine's "Exxon's Big Bet on Shale Gas." Further details, including Tillerson's quote, come from chapter 27 in Steve Coll's book on Exxon Mobil, cited below.

Details of Chesapeake's entry into the Barnett come from various sources, including its SEC filings (10-K in March 2002, as well as 8-Ks on March 12 and 18, 2002.) Hallwood's large wells in Johnson County were alluded to in a Morgan Stanley investor report by Lloyd Byrne on April 16, 2004. He refers to an unnamed private operator, but the details he provided match Hallwood wells in state records. The enthusiasm—and skullduggery—of the early Barnett days comes through in the *Noel v. Rees-Jones* lawsuit in state district court in Harris County (Houston, Texas.)

Nichols's decision not to join the Oklahoma City group that bought the Seattle SuperSonics can be found in a 2008 deposition given by McClendon as part of the federal court case *City of Seattle v. The Professional Basketball Club LLC* (2:07-cv-01620-MJP, Western District of Washington).

Burke, Bob. *Deals, Deals, and More Deals: The Life of John W. Nichols.* Oklahoma City: Oklahoma Heritage Association, 2004.

Coll, Steve. *Private Empire: ExxonMobil and American Power.* New York: Penguin Press, 2012.

Donnelly, John. "Q&A: J. Larry Nichols, CEO, Devon Energy." *Journal of Petroleum Technology* 58, no. 7 (July 2006): 30–31.

Ford, Charles, and Bob Burke. *The Oklahoma State Capitol: A History of Our Seat of Government.* Oklahoma City: Oklahoma Heritage Association, 2011.

Gold, R. "Boom Town: Drilling for Natural Gas Faces a Sizable Hurdle: Fort Worth—The U.S.'s Largest Field Lies Under 1.6 Million People; Not Everyone Will Benefit—Avoiding Baseball Diamonds." *Wall Street Journal*, April 29, 2005.

Nichols, J. Larry. *Devon Energy Corporation / J. Larry Nichols.* Exton, PA: Newcomen Society of the United States, 1996.

Siebrits, E., J. L. Elbel, R. S. Hoover, et al. "Refracture Reorientation Enhances Gas Production in Barnett Shale Tight Gas Wells." Paper presented at SPE Annual Technical Conference and Exhibition, Dallas, October 1–4, 2000.

Williams, Peggy. "Barnett Wonderland." *Oil and Gas Investor*, April 29, 2005.

Chapter 8: The Rise of Aubrey McClendon

I attended the 2012 Chesapeake shareholders' meeting, and quotes and details of that day are from my notebooks. An overview of the Sarbanes-Oxley Act, as it relates to audit committee, can be found in online guidelines, *Standards Relating to Listed Company Audit Committees*, on the US Securities and Exchange Commission website at www.sec.gov/rules/final/33-8220.htm. (Last accessed August 2013.) The historically low vote tally for directors is from Institutional Shareholder Services, a consulting firm. Ed Crooks, the *Financial Times*'s energy reporter in the United States, sat behind me in the reporter's quarantine at the meeting and deserves credit for noticing the Billie Holiday song that played after the meeting wrapped up. The steep $3 billion loan I mentioned was disclosed by the company in a May 11 press release. The "Yes, it's a crisis" quote is from Jon Wolff's note to investors, "Pricey $3B Loan Averts Liquidity Crisis," on May 14, 2012. He works for financial advisory firm International Strategy & Investment.

Nominal oil prices hit $39 a barrel in February 1981, equal to $103 in 2013 dollars. The source for this is the Energy Information Administration's "Real Prices Viewer" in the Short-Term Energy Outlook available here: www.eia.gov/forecasts/steo/realprices. (Last accessed August 2013.) The Robert Kerr campaign quote is from an exhibit at the University of Oklahoma's Carl Albert Center (http://web.archive.org/web/20070806164902/http://ou.edu/special/albertctr/archives/kerr/KERRPN3.HTM). (Last accessed August 2013.) Consultant WTRG Economics offers a useful guide to historical oil prices here: www.wtrg.com/prices.htm. (Last accessed August 2013.)

I have spoken with Ralph Eads, Tom Ward, and Aubrey McClendon many times over the past few years, as well as others at Randall & Dewey and at Jefferies. I have also interviewed Blair Thomas at EIG, Floyd Wilson, formerly of Petrohawk Energy, and various Chesapeake board members. The overwhelming majority of my interviews were on the record. On a few occasions, people wanted to speak without attribution. I agreed to this very warily, because I do not want to give people the ability to tell their story unless they are willing to attach their names. I have used these off-the-record interviews very sparingly. Jefferies's *U.S. Shale/Resource Play Overview*, a two-volume pitch book used to interest large foreign and institutional investors, was a helpful resource as well. The book changed over time, and the version I consulted was from April 2012.

If you want to learn more about the collapse of the Penn Square Bank, consult two sources: Mark Singer's *Funny Money* and part II, chapter 3 of *Managing the Crisis: The FDIC and RTC Experience 1980–1994*, published by the Federal Deposit Insurance Corporation in August 1998. Jerry Shottenkirk's profile of McClendon, cited below, is the source of McClendon's quote about spending time with his father "looking at dirty bathrooms in gas stations." Details of his time at Jaytex and Duke come from several sources, including interviews and McClendon's deposition in the *City of Seattle v. The Professional Basketball Club LLC*, as well as from Rick Robinson's 2002 profile, cited below.

Details about the early days of Chesapeake come from many sources, most notably a lengthy author interview with Tom Ward. McClendon has told parts of this story several times in hometown newspaper interviews. Please see Chapter 9 notes for more information about the sources of McClendon quotes. The "We clearly could not outthink a geologist or an engineer" quote comes from an industry conference on April 21, 2009. I pulled the court file from the Plotner lawsuit out of storage and read through the voluminous transcripts, depositions, and arguments. The case is no. 81896, *Tom L. Ward, TLW Investments, Aubrey McClendon, Chesapeake Investments and Chesapeake Operating Inc. v. Ralph E. Plotner Oil & Gas Investments Inc.*, and includes records from the jury trial. The quotes from Megan Hann, the court findings and related materials all come from the case file. McClendon contended that he and Ward were "100 percent" innocent, but did not provide any information to dispute the contemporary court record.

The S-1 form that Chesapeake filed with the SEC on February 4, 1993, is a fascinating document and details the birth of the Founders Well Participation Program. It also is where I found that Arthur Andersen had resigned as the company's independent accounting firm. The S-1 form is a lengthy form that all companies must file before selling stock to the public for the first time. It provided information about Ward's and McClendon's salaries

($175,000). The company reported capital expenditures of $37.6 million in its 1994 fiscal year. I calculated what 2.5 percent of that would be ($940,000). This is the source of my contention that McClendon needed to pay several times his annual salary. McClendon, during a talk with investors on November 16, 2004, remarked that Chesapeake's IPO was the worst performer of the year. The company's 1993 annual report is the source of the quote that Chesapeake "believes a financing arrangement of this type is unprecedented in the industry." While digging through Oklahoma Uniform Commercial Code (UCC) filings related to debt and sales, I discovered that four of the original board members lent McClendon and Ward $1.65 million. The key document was filed on August 24, 1993, and assigned no. 00111395 and was recorded in book 6,478, pages 0184–0199. My thanks to Theo Francis—an SEC filings investigative maven who ran his own consulting and research outfit, Disclosure Matters—for helping me decipher what Chesapeake was saying.

The source of Eads's quote about American Energy Operations comes from a 1986 *Wall Street Journal* article, "New Wildcatters: Oil-Patch 'Bargains' Lure Some Plungers Hoping to Get Rich," by Steve Frazier and James Tanner. My claim that there is one oil operator for every 1,200 Oklahomans—twice the rate of Texas—comes from dividing the state's population by the number of active operators in May 2012 reported by the Oklahoma Corporation Commission. I did the same with Texas Railroad Commission records. Thanks to Ken Zimmerman, a longtime employee of the Oklahoma Corporation Commission, for his insights on the nature of the Oklahoma wildcatter. David Fleischaker, the state's energy secretary from 2003 to 2008, also shared his recollections. I first reported some details of Chesapeake's near-death experience in the Austin Chalk and meeting with Calpine in a 2006 front-page article in the *Wall Street Journal*, and fleshed out details for the book by looking through news articles and corporate filings.

Getting to the bottom of the California energy crisis and El Paso's and Ralph Eads's role required a lot of reading and digging through Federal Energy Regulatory Commission dockets. Some of the most interesting documents have been made public only in recent years and were not available to reporters covering the story as it unfolded. The March 2003 FERC final report, cited below, is the best overview document. El Paso fought to keep the Eads memo to Bill Wise confidential, but it was eventually made public and can be found in slides 19–21 of the December 13, 2002, filing in *Public Utilities Commission of the State of California, et al., v. El Paso Natural Gas Company, et al.*, FERC Docket no. RP00-241-006. Also informative was Administrative Law Judge Curtis Wagner's initial decision, issued on September 23, 2002, and part of the above docket. For details on power and gas prices, see *Public Utilities Commission of the State of California v. El Paso Natural Gas Company, et al.,* FERC Docket no. RP00-241-000. California's initial brief, filed on August 24, 2001, is particularly instructive. The "cashed in big-time . . ." quote is from a transcript of a FERC hearing on December 2, 2002, and was spoken by Harvey Morris. He, Wagner, and Frank Lindh, an attorney for Pacific Gas and Electric Company, all shared their recollections of the clash in the Washington courtroom. I also relied on coverage in the *Wall Street Journal* and the *Houston Chronicle* for details of the El Paso settlement and investigation of price reporting. El Paso reported the restatement of its earnings in an 8-K (August 23, 2004) and a 10-K (September 30, 2004) filed with the SEC. FERC's settlement order was issued on November 14, 2003, and is filed in Dockets nos. RP00-241-000 and RP00-241-006. Details of El Paso's internal investigation into index price reporting is in FERC's final report, cited below.

Driver, Anna, and Brian Grow. "The Energy Billionaire's Shrouded Loans." Reuters, April 18, 2012.

Federal Energy Regulatory Commission. *Final Report on Price Manipulation in Western Markets: Fact-Finding Investigation of Potential Manipulation of Electric and Natural Gas Prices, Docket No. PA02-2-000.* Washington, DC: Federal Energy Regulatory Commission, March 2003.

Gold, R. "Investment Forecast: On a Roller Coaster, One Energy Firm Tries Hedging Bets; Natural-Gas Giant Chesapeake Has Expanded Rapidly by Locking In Its Revenue; Dangers of Guessing Wrong." *Wall Street Journal,* November 6, 2006.

Gold, R., and Daniel Gilbert. "Chesapeake Directors Spurned: Two Board Members Fall Far Short of Voting Majority in Rebuke to Gas Driller." *Wall Street Journal,* June 9, 2012.

Gold, R., Daniel Gilbert, and Joann S. Lublin. "For Chesapeake's CEO, a Complex Web of Loans." *Wall Street Journal,* April 19, 2012.

Krauss, Clifford, and Eric Lipton. "After the Boom in Natural Gas." *New York Times,* October 21, 2012.

McLean, Bethany, and Peter Elkind. *The Smartest Guys in the Room: The Amazing Rise and Scandalous Fall of Enron.* New York: Portfolio, 2003.

Mollenkamp, Carrick. "Chesapeake's Deepest Well: Wall Street." Reuters, May 9, 2012.

Robinson, Rick. "The Daily Oklahoman Executive Interview Column." *Daily Oklahoman*, July 21, 2002.

Shottenkirk, Jerry. "Hard Work, Luck Make Billions for Oklahoma Executive." *Journal Record* (Oklahoma), August 13, 2007.

Smith, Rebecca, and John R. Emshwiller. *24 Days: How Two Wall Street Journal Reporters Uncovered the Lies That Destroyed Faith in Corporate America*. New York: HarperCollins, 2003.

Chapter 9: The Fall of Aubrey McClendon

I had spoken to and exchanged emails with Aubrey McClendon for many years and wrote a front-page profile of him and Chesapeake for the *Wall Street Journal* in 2006, back when he was known only inside energy circles. I had scheduled a lengthy interview with him on May 1, 2012, to go over his history and the story of Chesapeake and natural gas. About ten days before the interview, Reuters reporters Anna Driver and Brian Grow first reported the existence of enormous loans to McClendon from financial firms that were buying assets from Chesapeake. Their article, "The Energy Billionaire's Shrouded Loans," is available online at http://graphics.thomsonreuters.com/12/04/ChesapeakeMcClendon.pdf. (Last accessed August 2013.) My follow-up story on April 19 provided some additional details and is the source of the $1.4 billion figure.

McClendon canceled my May 1 interview and subsequently declined to reschedule it. He responded to some questions I sent to him by email during the editing of this book. His responses are the source of several quotes attributed to him in Chapters 8, 9, and 11. Before April 2012, he was never shy about talking at conferences and giving interviews. I compiled all the transcripts of his talks on investor conference calls and at energy conferences, and it is not an exaggeration when I say that the file ran to nearly a thousand pages. I also relied on tapes of speeches he gave, lengthy emails we exchanged in 2006, and tape recordings of interviews he gave to other reporters.

His quote "It will be better to be a provider of energy than a consumer of energy for the next twenty years" comes from an article in the November 4, 2002, edition of the *Oil & Gas Journal,* and his pessimism about future production is from the March 24, 2003, edition of the same publication. The *OGJ* publishes an annual list of the largest oil and gas producers. I relied on these rankings to make the determination that Chesapeake was the largest driller in the country.

Jason Baihly, in a 2010 paper cited below, provides detail on how shale wells grew larger and the industry figured out how to drill them more quickly.

A quick note on company financials. The $328.8 million figure in the second quarter of 2004 is cash from operations; the $337.9 million is a broad definition of capital expenditures. These figures are derived from S&P Capital IQ, a financial data product. From that point until the end of 2012, Chesapeake reported three quarters of positive cash flow: the first quarter of 2005 and the second and third quarters of 2006. The $163 billion invested in US shale is roughly equal to the market capitalization of Coca-Cola ($167.6 billion) and Google ($165.4 billion). Those figures are from the 2012 FT Global 500 listing, published by the *Financial Times* and current as of March 30, 2012.

This chapter deals, at length, with McClendon's loans. Details on those loans are contained in UCC filings in Oklahoma City. His ties with John Arnold's Centaurus Advisors are in UCC filings nos. 2008011942027 and 2008013852736. See also filings nos. 20100517020492600 and 2008008596947, although this is not an exhaustive list. Chesapeake disclosed information about McClendon's stake in the Founders Well Participation Program in a SEC DFAN14A filing on April 26, 2012. I consulted the Energy Information Administration's *Natural Gas Monthly* reports for my assertion that McClendon's 147 million cubic feet a day of natural was enough to supply all residential customers in Connecticut or Kentucky. See table 14 of the monthly EIA reports. For more information about Arnold, see "The Reckoning of Centaurus Billionaire John Arnold," by Leah McGrath Goodman in the February 1, 2011, edition of *Absolute Return*.

Some of McClendon's quotes about Chesapeake having four main inputs and the number of leasing transactions come from an interview he gave on January 5, 2012, to my *Wall Street Journal* colleagues Daniel Gilbert and Ryan Dezember.

Senator Bernie Sanders, in August 2011, released a snapshot of participants in natural gas and oil futures markets. The information was made available on his Senate website, www.sanders.senate.gov/newsroom/news/?id=e802998a-8ee2-4808-9649-0d9730b75ea4. (Last accessed August 2013.) I downloaded the data, put it into spreadsheets, and analyzed it to come up with the rankings of largest natural gas market participants. In a statement on August 25, 2011, the Futures Industry Association said it was "shocked and outraged" by the disclosure.

Details on Chesapeake's $8 billion in gains from trading and hedging were reported by the company over time in various public disclosures. The section on the Heritage Management hedge fund is based on reporting by myself and the former *Wall Street Journal* reporter Ann Davis in 2006 and 2007 and was supplemented by the Reuters article cited below.

John Pinkerton's quote "When Aubrey joined the party" is from an interview with the author. McClendon's quote "I got caught up in a wildfire" is from Ben Casselman's *Wall Street Journal* article cited below. I greatly appreciate Casselman's insights from covering Chesapeake for the *Journal* in this time frame. The value of McClendon's share ownership is derived from various SEC filings.

More broadly, I have benefited from many, many talks over the years with investment bankers and analysts who followed Chesapeake. I quote Dan Pickering in the chapter, but he is not alone.

Pawel Rajszel's investor note was titled "Chesa'peake Leverage: It's More Than You Think" and issued by Veritas Investment Research on April 29, 2010. McClendon's quote that regarding VPPs as debt is "kind of nutty" is from a conference call with analysts on November 4, 2010. His appearance on *Mad Money* was archived on Chesapeake Energy's website at www.chk.com/News/Articles/Pages/TV_20110801_AKM.aspx. (Last accessed August 2013.)

The Archie Dunham email was sent on January 29 and reviewed by the author. McClendon's email in April 2013 was first reported by Christopher Helman in *Forbes* in an article titled "Aubrey McClendon Is Now Hiring."

Baihly, Jason, Raphael Altman, Raj Malpani, and Fang Luo. "Shale Gas Production Decline Trend Comparison over Time and Basins." Paper presented at SPE Annual Technical Conference and Exhibition, September 19–22, 2010, Florence, Italy.

Birol, Fatih. *World Energy Outlook 2012*. Paris: IEA Publications, 2012.

Casselman, Ben. "Margin Calls Hitting More Executive Suites." *Wall Street Journal*, October 13, 2008.

Gold, R. "Costly Liabilities Lurk for Gas Giant." *Wall Street Journal*, May 11, 2012.

Gold, R., and Daniel Gilbert. "The Many Hats of Aubrey McClendon." *Wall Street Journal*, May 8, 2012.

Schneyer, Joshua, Jeanine Prezioso, and David Sheppard. "Special Report: Inside Chesapeake, CEO Ran $200 million Hedge Fund." Reuters, May 2, 2012.

Chapter 10: Celestia

My visit to Sullivan County was greatly enriched by the time I spent with innumerable local residents and community leaders. At the suggestion of former *Pittsburgh Post-Gazette* reporter Erich Schwartzel, I convinced Commissioner Bob Getz to give me the grand tour of the county. The tour didn't disappoint. Many of the people I interviewed are in this chapter, but I would specifically like to thank Dick and Lois McCarthy—former farmers, neighbors of the Farm, and local leaders. And thanks to Dean Homer, who serves as both the county's only undertaker and one of its only accountants. When it comes to death and taxes, see Dean Homer. Ralph Kisberg helped with insights and logistics. Andy Goldberg, a Harvard-trained Wall Street trader turned independent water tester in nearby Clarks Summit, helped me deepen my understanding of water. Tom Murphy, codirector of the Penn State Marcellus Center for Outreach and Research, is a state treasure.

Geological information about Sullivan County came from an old (and undated) pamphlet titled *Worlds End State Park, Geologic Features of Interest*, published by the Bureau of Topographic and Geologic Survey. Geoscientist Fred Baldassare, who has studied the commonwealth's rocks for decades, answered my many questions.

The Reibsons' court case is Chesapeake Appalachia LLC v. Milo K. and Betty Reibson, 4:11-cv-01321-TMB. The lawsuit was settled on November 5, 2012. On November 13 Chesapeake applied for a well permit for the Phillips Unit. The site ID in the state's Department of Environmental Protection computer system is 739392.

My count of wells in Sullivan and nearby counties was based on the Pennsylvania Department of Environmental Protection Office of Oil and Gas Management's *Well Inventory by County Report*, retrieved in August 2012. I counted only "unconventional wells." Details on the Marc 1 Hub pipeline can be found in the Federal Energy Regulatory Commission Docket no. CP10-480-000. Especially helpful was the *Environmental Assessment*, filed in May 2011, and a PowerPoint presentation by Inergy, the company that built the pipeline, on June 28, 2010, and is titled "Project Introduction Meeting with FERC Staff on CNYOG's MARC 1 Hub Line."

Information about coal plant retirements is from the Energy Information Administration's *Annual Energy Outlook 2012*. MIT's *The Future of Natural Gas* report presents a concise overview of the impact of the Powerplant and Industrial Fuel Use Act of 1978. Ronald Reagan's quote can be found here: www.presidency.ucsb.edu/ws/index.php?pid=34320#axzz1W3ISgYEl. (Last accessed August 2013.)

There have been numerous reports on the wealth generated by shale exploration. I recommend "Oil Boom in Eagle Ford Shale Brings New Wealth to South Texas" by Robert W. Gilmer, Raúl Hernandez, and Keith R. Phillips, published in the Second Quarter 2012 edition of the Federal Reserve Bank of Dallas publication *Southwest Economy*. The gas industry has paid for other reports, including *The Economic and Employment Contributions of Unconventional Gas Development in State Economies* by Mohsen Bonakdarpour and John W. Larson, IHS, published in June 2012. See also *America's New Energy Future: The Unconventional Oil and Gas Revolution and the US Economy* by John W. Larson, Richard Fullenbaum, Richard Slucher, et al., IHS, October 2012.

John Pinkerton, the former CEO, and Ray Walker were both very forthcoming about both the missteps and successes of Range Resources's early involvement in the Marcellus and the difficulties in getting any modern equipment. Walker was the befuddled consultant who made plans to go to Pittsburg, Texas. The company has talked about the Renz #1 well, as well as the number of acres it had leased, on conference calls with analysts. Chesapeake talked about its first Marcellus well in a conference call with Wall Street analysts on August 3, 2007.

I spoke with both former Pennsylvania governor Ed Rendell and Ed Cohen, former CEO of Atlas Energy, at the *Wall Street Journal*'s ECO:Nomics conference in March 2012. This is the source of Cohen's "farmers" quote. Rendell is the source of the statistic that showed the number of well permits to drill in the Marcellus Shale skyrocketing between 2007 and 2010.

The state investigation of the June 2010 blowout of an EOG well in Clearfield County, Pennsylvania, is remarkably thorough. Bedrock Engineering, under contract, produced a *Well Control Incident Analysis* report in July 2010 that included transcripts of interviews with people involved in drilling the well.

I obtained many documents about drilling in Pennsylvania through Right to Know Law requests. This is the source of the section on Chesapeake's admission that its wells intersected a fault. I also obtained records related to Shell's blowout in Tioga County.

The source of McClendon's "problem identified" quote is from the September 2011 Marcellus Shale Insights Conference. It is archived here: www.chk.com/news/articles/pages/20110911_akm.aspx. (Last accessed January 2013.)

I did not attend the Engelder-Ingraffea debate in Laporte but was able to watch it on YouTube. Here are the relevant videos, in order: www.youtube.com/watch?v=Di1JKZEbHyQ; www.youtube.com/watch?v=1u6aqXaI3s4;

www.youtube.com/watch?v=kCYUdLJSav8; and www.youtube.com/watch?v
=doA5-PCG530. (Last accessed August 2013.)

Bender, D. Wayne, *From Wilderness to Wilderness: Celestia.* Laporte, PA: Sullivan County Historical Society, 1995.

Boyer, Elizabeth W., Bryan R. Swistock, James Clark, Mark Madden, and Dana E. Rizzo. *The Impact of Marcellus Gas Drilling on Rural Drinking Water Supplies* Harrisburg, PA: Center for Rural Pennsylvania, 2012, updated.

Brown, Robbie. "Gas Drillers Asked to Change Method of Waste Disposal." *New York Times,* April 20, 2011.

Brown, Sara. Unpublished and untitled senior thesis, Elizabethtown College, 2006, available at Sullivan County Historical Society and Museum.

Gold, R. "Faulty Wells, Not Fracking, Blamed for Water Pollution." *Wall Street Journal,* March 13, 2012.

Klein, Philip Shriver, and Ari Arthur Hoogenboom. *A History of Pennsylvania.* New York: McGraw-Hill, 1973.

Legere, Laura. "Stray Gas Plagues NEPA Marcellus Wells." *Daily Review* (Towanda, PA), July 10, 2011.

McGraw, Seamus. *The End of Country.* New York: Random House, 2011.

Silver, Jonathan D. "Origins: The Story of a Professor, a Gas Driller and Wall Street." *Pittsburgh Post-Gazette,* March 20, 2011.

Waples, David A. *The Natural Gas Industry in Appalachia: A History from the First Discovery to the Maturity of the Industry.* Jefferson, NC: McFarland, 2005.

Wilber, Tom. *Under the Surface: Fracking, Fortunes and the Fate of the Marcellus Shale.* Ithaca, NY: Cornell University Press, 2012.

Chapter 11: Blessings of the Pope
While some pieces of this chapter have been reported elsewhere, the full story of Aubrey McClendon and Carl Pope has never been told before. I am grateful to Pope, Michael Brune, and other Sierra Club officials, including Ken Kramer, Jeremy Doochin, Barbara Frank, and Roger Downs, for sharing their recollections. Alan Nogee and Tim Wirth talked to me about environmentalists and their response to climate change and natural gas. Scott Anderson at the Environmental Defense Fund has, over many years, helped me find my way through confusing and contradictory statements by the industry and environmentalists.

Details of the 2011 protest march came from participants as well as from uploaded YouTube videos. The source of McClendon's "We're cold, it's dark, and we're hungry" quote is the September 2011 Marcellus Shale Insights Conference. It is archived here: http://www.chk.com/news/articles/pages/20110911_akm.aspx. (Last accessed August 2013.)

I reviewed copies of the *Houston Chronicle*, *Austin American-Statesman*, and other Texas newspapers on microfiche at the Austin Public Library to find the "Coal Is Filthy" ads. These ads were well covered by R. G. Ratcliffe at the *Chronicle* and Asher Price at the *American-Statesman*. Their articles helped piece together the broader story. The Tom Price quote is from an *American-Statesman* article. Texas secretary of state records indicate that the Clean Sky Coalition was registered on January 26, 2007, and that it was formed as Delaware nonprofit corporation a week earlier, on January 19. The coal lobby's "disturbing departure" quote is from an April 22, 2008, *Washington Post* article, "Breaking the 'Be Nice' Rule in the Energy Family."

Pope's "Among the fossil fuels . . ." quote is from Nissa Darbonne's interview with him, "An Unexpected Union."

The history of TXU came from corporate material and the US Interior Department's announcement that it was expanding the boundaries of the Dallas Downtown Historic District. Its $10 billion coal plan was covered in specialist publications such as *Megawatt Daily* as well as in the *Dallas Morning News*. More details are in the "background of the merger" section of TXU's Prem-14a SEC filing on June 14, 2007. Details of the electrical power generation are from the Energy Information Administration's *Electric Power Annual 2006*. Data on natural gas surpassing coal in Oklahoma and equaling coal output nationwide in April 2012 are from table 1.1 of the Energy Information Administration publication *Electric Power Monthly*. I also used this monthly publication for the changing US electrical mix. The EIA's *Monthly Energy Review* was the source for my discussion of falling carbon dioxide emissions; see table 12.1. It must be noted that the sluggish economy, as well as coal-to-gas switching and improved energy efficiency, played a role in lower emissions.

Details of the February 2007 rally can be found in the February 16, 2007, edition of the *Austin Chronicle*, as well as a report and photographs on the *Burnt Orange Report*, a blog devoted to Lone Star politics. It can be found here: www.burntorangereport.com/diary/2890. (Last accessed August 2013.)

McClendon hasn't talked at length about climate change, but his comments to a conference held by the Independent Petroleum Association of America on October 3, 2006, provided quotes and a degree of clarity. His "Coal is the wrong answer for Oklahoma today" quote is from a July 31,

2007, article in the *Daily Oklahoman*. His letter in the *Tulsa World*—the source of his quote that "Coal is simply on the wrong side of history"— appeared on August 26, 2007. "What's the matter with self-interest? . . ." was in a *Fort Worth Star-Telegram* article on September 27, 2007. "We crave volatility" comes from a panel he appeared on at CERAWeek in 2007. The Thomas Friedman column, titled "Global Weirding Is Here," appeared in the *New York Times* on February 17, 2010.

I read several years' worth of Carl Pope's bimonthly columns in *Sierra*, the Sierra Club's magazine. He wrote about the threat posed by global warming to "all our previous gains" in "A World Transformed" in the March–April 2007 issue. I also read many issues of the Pennsylvania chapter's newsletter, the *Sylvanian*, and the New York chapter's *Sierra Atlantic*. The *Alameda* magazine profile of Brune, cited below, is the source of the story of his takeover of the PA system at Home Depot. Pope's claim, "We will not take money from oil companies," comes from a speech he gave at the City Club of Cleveland in June 2008. It is viewable here: http://www.youtube.com/watch?v=jQBkhXDAZKk. (Last accessed August 2013.) The Sierra Club's *Energy Resources Policy* was first published in September 2006 and then updated in May 2009 and July 2011.

To learn more about the Ross #1, I reviewed filings by Gastem, the New York State legislature, and the New York Department of Environmental Conservation. The December 4, 2011, *New York Times* published a profile of actor Mark Ruffalo that included his neighbor's quote, under the headline "Ruffalo Embraces a Role Closer to Home."

The IEA data on the impact of increased gas consumption on carbon dioxide emissions that Michael Brune discussed can be found on page 91 of the IEA publication *Golden Rules for a Golden Age of Gas*, which is cited below. The quote about how people "may find it difficult to accept" that gas helps slow the rise in carbon dioxide is on page 99 of this report.

The Citigroup report I referenced is called *Shale & Renewables: A Symbiotic Relationship*. It was written by Jason Channell and issued on September 12, 2012.

Birol, Fatih. *World Energy Outlook 2011. Special Report: Are We Entering a Golden Age of Gas?* Paris: IEA Publications, 2011.

———. *Golden Rules for a Golden Age of Gas*. Paris: IEA Publications, 2012.

Brune, Michael. *Coming Clean: Breaking America's Addiction to Oil and Coal*. San Francisco: Sierra Club Books, 2008.

Casselman, Ben. "Sierra Club's Pro-Gas Dilemma: National Group's Stance

Angers On-the-Ground Environmentalists in Several States." *Wall Street Journal*, December 21, 2009.

Cohen, Michael P. *The History of the Sierra Club, 1892–1970*. San Francisco: Sierra Club Books, 1988.

Darbonne, Nissa. "Dear Sierra Club, Do You Take This U.S. Natural Gas Industry? An Unexpected Union." *Oil and Gas Investor*, March 4, 2008.

Devall, Bill. "The End of American Environmentalism?" *Nature and Culture* 1, no. 2 (Autumn 2006): 157–80.

Freeman, James. "The Weekend Interview with Aubrey McClendon: The Politically Incorrect CEO." *Wall Street Journal*, April 28, 2012.

Intergovernmental Panel on Climate Change. *IPCC Special Report on Renewable Energy Sources and Climate Change Mitigation*. Cambridge, UK: Cambridge University Press, 2011.

Kennedy Jr., Robert F. "How to End America's Deadly Coal Addiction." *Financial Times*, July 19, 2009.

Leaton, James. *Unburnable Carbon—Are the World's Financial Markets Carrying a Carbon Bubble?* London, UK: Carbon Tracker Initiative, 2011.

McKibben, Bill. "Global Warming's Terrifying New Math." *Rolling Stone*, July 19, 2012.

Pope, Carl. *Sahib: An American Misadventure in India*. New York: Liveright, 1972.

Pope, Carl, and Paul Rauber. *Strategic Ignorance: Why the Bush Administration Is Recklessly Destroying a Century of Environmental Progress*. San Francisco: Sierra Club Books, 2004.

Robbins, Noelle. "Speak Softly and Carry a Big Green Stick: Michael Brune Kicks Butt for the Rainforest Action Network." *Alameda*, July–August 2009.

Shogren, Elizabeth. "Natural Gas as a Climate Fix Sparks Friction." NPR *Morning Edition*, February 23, 2010.

Walsh, Bryan. "How the Sierra Club Took Millions from the Natural Gas Industry—and Why They Stopped." *Time*, February 2, 2012.

Chapter 12: Ghost Ridin' Grandpa

The YouTube video of Claude Cooke and his wife ghost riding is available at www.youtube.com/watch?v=SBPJTK1YBvs. When I last watched it, in August 2013, it had nearly 2.8 million views. Cooke first contacted me after my "Faulty Wells, Not Fracking, Blamed for Water Pollution" article appeared on the front page of the *Wall Street Journal* in March 2012.

The source of the estimation that about $5 billion is spent on cementing, out of a total outlay of $105 billion, comes from correspondence with Richard Spears of Spears & Associates. He admits that determining the actual spending is difficult and the $5 billion figure should be treated as an informed estimate.

Details on the sixteen families whose properties were bought by Chesapeake can be found in the federal court case *Phillips et al. v. Chesapeake Appalachia LLC* in the Middle District of Pennsylvania. Also, the state Department of Environmental Protection press release "DEP Fines Chesapeake Energy More Than $1 Million," dated May 17, 2011.

I have met and talked with Claude Cooke—and his grandson Brian Smiley—numerous times. The biographical material about Cooke came from interviews. The American Petroleum Institute's calling Cooke's work a "revelation" is from API Recommended Practice 65—Part 2, *Isolating Potential Flow Zones During Well Construction* (1st ed., May 2010). API Technical Report 10TR1, "Cement Sheath Evaluation" (2nd ed., September 2008), provides a thorough account of cement-testing technology and praises the isolation scanner, though it notes that the tool's advanced sensor capability "does not probe deeper into the cement sheath" (section 9.8.5).

I constructed Marvin Gearhart's biography largely from an 1982 oral history interview by Floyd Jenkins, available through the University of North Texas Oral History Program, Denton, Texas. This was the source of information about Gearhart Industries and the quote "I got it in my blood." I also relied on contemporary newspaper accounts of the merger with Halliburton, published in the *Dallas Morning News* and Fort Worth *Star-Telegram*, and Halliburton filings with the SEC.

Mark Zoback's quote "There are three keys—and those are well construction, well construction, and well construction" is from an interview and first appeared in my March 2012 *Wall Street Journal* article.

The radial differential temperature tool was issued US Patent no. 4,074,756 and the related US Patent no. 4,109,717. In addition, I relied on Cooke's publications, the most important of which are cited below.

Information about the role of cement and the failed negative pressure test is from numerous sources compiled by the National Commission on the BP Deepwater Horizon Oil Spill and Offshore Drilling, including the chief counsel's report, which went into detail about cementing. The quote "cementing an oil well . . ." is from page 99 of the report *Deep Water: The Gulf Oil Disaster and the Future of Offshore Drilling—Report to the President*. The detail about BP ordering well integrity testing is on page 95 of the chief counsel's report *Macondo: The Gulf Oil Disaster*. Three key reports are cited below. Further information about what happened was first reported by my colleague Ben Casselman and me in the May 27, 2010, *Wall Street Journal* article "Unusual Decisions Set Stage for BP Disaster."

Information about the Flatirons well comes from the "Evaluation of Precompletion Annular Gas Leaks in a Marcellus Lateral" paper cited below and from conversations with Jeff Jones, a managing director at the company, and Bob Ging, an attorney who represented Brockway Borough. Further details are from Pennsylvania Department of Environmental Protection internal reports and documents obtained through Right to Know Law requests. The well is DU 3-6-1H in Snyder, Pennsylvania. Among the papers I obtained that were important were daily inspection reports; a summary of a Section 501 conference meeting held on August 3, 2011; a Flatirons PowerPoint presentation dated February 23, 2011; and a Flatirons presentation discussing the pros and cons of remedial cementing, dated July 27, 2011.

Everything you ever wanted to know about cow emissions of natural gas can be found in the study "Model for Estimating Enteric Methane Emissions from United States Dairy and Feedlot Cattle," cited below.

Information about US natural gas usage is from *Trends in U.S. Residential Natural Gas Consumption*, published in June 2010 by the Energy Information Administration. I read several papers and Schlumberger publications on its isolation scanner, including "Ensuring Zonal Isolation Beyond the Life of a Well" by Mario Bellabarba, Hélène Bulte-Loyer, Benoit Froelich, et al. in the Spring 2008 issue of *Oilfield Review*.

Bartlit Jr., Fred H. *Macondo: The Gulf Oil Disaster—Chief Counsel's Report*. Washington, DC: US Government Printing Office, 2011.

Bureau of Ocean Energy Management, Regulation, and Enforcement. *Report Regarding the Causes of the April 20, 2010, Macondo Well Blowout* (September 2011). Joint Marine Board report by US Coast Guard and BOEMRE.

Cooke, Claude E. "Fracturing with a High-Strength Proppant." *Journal of Petroleum Technology* 29, no. 10 (October 1977): 1222–26.

———. "Radial Differential Temperature (RDT) Logging—A New Tool for Detecting and Treating Flow Behind Casing." *Journal of Petroleum Technology* 31, no 6 (June 1979): 676–82.

Gold, R., and Ben Casselman. "Cementing, Mainstay of Oil Drilling, Is Prone to Failure." *Wall Street Journal*, October 30, 2010.

Legends of Drilling (CD-ROM). Houston: Society of Petroleum Engineers, 2009.

Legends of Hydraulic Fracturing (CD-ROM). Houston: Society of Petroleum Engineers, 2010.

Meyer, Andre J., and Claude E. Cooke Jr. "Application of Radial Differential Temperature (RDT) Logging to Detect and Treat Flow Behind Casing." Paper presented at SPWLA 20th Annual Logging Symposium, June 3–6, 1979, Tulsa.

Moore, L. P., J. E. Jones, S. H. Perlman, and T. A. Huey. "Evaluation of Pre-completion Annular Gas Leaks in a Marcellus Lateral." Paper presented at SPE Americas Unconventional Resources Conference, June 5–7, 2012, Pittsburgh.

National Commission on the BP Deepwater Horizon Oil Spill and Offshore Drilling. *Deep Water: The Gulf Oil Disaster and the Future of Offshore Drilling—Report to the President.* Washington, DC: US Government Printing Office, 2011.

Chapter 13: Pandora's Frack

Most of the details of the Perot Museum of Nature and Science come from my visit a couple weeks after it opened. I also exchanged emails with Arne Emerson, a principal at Morphosis, the architecture firm that designed the building, about the inspiration of the design.

The "insurgency" quote is from a CNBC segment titled "Oil Executive: Military-Style 'Psy Ops' Experience Applied" by Eamon Javers and is available at www.cnbc.com/id/45208498. (Last accessed January 2013.)

The Stanolind researcher's use of hydrafrac treatment, see J. B. Clark, C. R. Fast, and G. C. Howard, "A Multiple-Fracturing Process for Increasing the Productivity of Wells" in *Drilling and Production Practice* (1952). I also learned about the history of the words *frak* and *frack* in *Battlestar Galactica* from Ron Moore, who won a Peabody Award in 2006 as a writer and producer. Chris Pappas, a collector of *Battlestar Galactica*–related materials, provided the 1978 writer's guide with the original spelling of *frack*. The Associated Press wrote a story about the word *frak* sweeping "geek nation" in 2008.

Details of the Southlake battle over fracking is mostly from local news reports, supplemented by interviews with participants and a review of Southlake files and legal documents. The Bartonville material comes from interviews with Ron Robertson and a review of its Ordinance 526-11 issued in October 2011. I also relied on the Texas Railroad Commission's "RRC Public GIS Map Viewer" and related documents.

A good overview of Germany's *Energiewende* can be found in Osha Gray Davidson's chapbook, cited below. Another fascinating article, and the source of the "fighting for its economic survival" quote, is "How to Lose Half a Trillion Euros" in the *Economist*'s October 12, 2013, edition. The Rex Tillerson quote "If you want to live by the precautionary principle . . ." is from an article in *Fortune* by Brian O'Keefe on April 16, 2012, titled "Exxon's Big Bet on Shale Gas."

I mentioned that the industry has more than doubled the number of wells drilled in a decade. The Energy Information Administration reports the number of oil and gas wells drilled in the United States as follows: 54,302 in 2008 and 22,911 in 1998. These wells were also longer, on average, according to EIA records of the total footage of wells drilled.

Information about Wyoming's ozone problems are briefly discussed in the Shale Energy Advisory Board's *Ninety-Day Report*, cited below. The report cites a July 2011 presentation from the Wyoming Department of Environmental Quality, which I also reviewed. The board's *Second Ninety-Day Report* discusses how wells can emit methane and other pollutants. The EPA, in August 2012, reported 0.078 part per million of ozone in Sublette County, Wyoming. Chicago and Phoenix reported 0.077 part per million. This information can be found at www.epa.gov/airtrends/values.html and clicking on the ozone information link. (Last accessed May 2013.) The Texas Commission on Environmental Quality benzene finding was from a Railroad Commission notice issued in March 2010 and available at www.rrc.state.tx.us/forms/reports/notices/airemission21010.pdf. (Last accessed August 2013.) Another source of information about air emissions from oil and gas operations is the June 2012 testimony by Environmental Defense Fund president Fred Krupp before the United States Senate Subcommittee on Clean Air and Nuclear Safety.

Information about the health impact assessment in Battlement Mesa, Colorado, was published in an *American Journal of Public Health* article titled "The Use of Health Impact Assessment for a Community Undergoing Natural Gas Development." The article was published online in April 2013. I also interviewed one of the authors, Lisa McKenzie, who said it was the first such assessment of shale-related development. The Health Impact Project, a collaboration between Pew Charitable Trusts and Robert Wood Johnson Foundation, tracks these assessments. Its website is www.healthimpactproject.org.

For a broad discussion of the macroeconomic and geopolitical implications of rising US crude oil production, please see the Morse publication cited below as well as Michael Levi's book, especially chapter 3.

The OPEC quote about the impact of the shale boom on US imports comes from the *Wall Street Journal* article "Beijing is Set to Overtake the U.S. as World's Largest Importer of Oil," by Benoît Faucon in the April 4, 2013, edition.

Alan Greenspan provided testimony to the Congressional Joint Economic Committee on November 3, 2005. Steve Coll, in *Private Empire*, cited above, provides evidence that Lee Raymond was Greenspan's tutor.

Details on the Nucor upgrader is from an article I wrote in the February 8, 2012, *Wall Street Journal* titled "Oil and Gas Boom Lifts U.S. Economy" and came from an interview with company president John Ferriola. Information about industrial investment in the United States is from Nancy Lamb at Dow Chemical. She provided me with the company's December 2012 version of a presentation it titled "Industry to Invest $90 Billion in Manufacturing Renaissance."

The Zaki Yamani quote is famous and can be found, among other places, in the *Economist*'s October 23, 2003, issue in an article titled "The End of the Oil Age." The "There is always either too much or too little" quote is from Paul H. Frankel, *The Essentials of Petroleum: A Key to Oil Economics,* 3rd ed. (London: Frank Cass, 1983). Thanks to Phil Verleger for helping me track down the quote—and for suggesting the Pandora's box metaphor.

At the time of this writing, there was a relatively small number of important studies on methane leakage. I have no doubt that more will follow. I relied on the Howarth letter—referred to as the Cornell study—and the article by O'Sullivan and Paltsev, cited below. In the O'Sullivan article, the supplemental materials sections S3 and S4 were particularly helpful.

Davidson, Osha Gray. *Clean Break: The Story of Germany's Energy Transformation and What Americans Can Learn from It.* New York: Inside Climate News, 2012.

Howarth, Robert W., Renee Santoro, and Anthony Ingraffea. "Methane and the Greenhouse-Gas Footprint of Natural Gas from Shale Formations: A Letter." *Climactic Change* 106, no. 4 (2011): 679–90.

Levi, Michael. *The Power Surge: Energy, Opportunity, and the Battle for America's Future.* New York: Oxford University Press, 2013.

Morse, Edward L., Eric G. Lee, Daniel P. Ahn, Aakash Doshi, Seth M. Kleinman, and Anthony Yeun. *Energy 2020: North America, the New Middle East?* Citi GPS: Global Perspectives & Solutions. http://fa.smithbarney .com/public/projectfiles/ce1d2d99-c133-4343-8ad0-43aa1da63cc2.pdf. (Last accessed April 2013.)

O'Sullivan, Francis, and Sergey Paltsev. "Shale Gas Production: Potential Versus Actual Greenhouse Gas Emissions." *Environmental Research Letters* 7, no. 4 (2012).

Secretary of Energy Advisory Board. *Shale Gas Production Committee Ninety-Day Report*. Washington, DC: August 2011.

———. *Shale Gas Production Committee Second Ninety-Day Report*. Washington, DC: November 2011.

Index

ABOUT THE AUTHOR

Russell Gold has reported on energy in *The Wall Street Journal* since 2002. His coverage of the Deepwater Horizon oil spill was honored with a Gerald Loeb Award and was a finalist for the Pulitzer Prize. He lives in Austin, Texas.

Visit russellgold.net for a teaching guide.